Perilous Knowledge

by the same author
BRITISH SCIENCE AND POLITICS SINCE 1945

TOM WILKIE

Perilous Knowledge

THE HUMAN GENOME PROJECT
AND ITS IMPLICATIONS

UNIVERSITY OF CALIFORNIA PRESS
Berkeley   Los Angeles

University of California Press 1993
Berkeley and Los Angeles, California

Published by arrangement with Faber and Faber Limited,
London, England

Libarary of Congress Cataloging-in-Publication Data

Wilkie, Tom.
Perilous knowledge: the human genome project and its implications/Tom Wilkie.
p.   cm.
Includes index.
ISBN 0-520-08553-1
1. Human Genome Project.   2. Human gene mapping—Moral and ethical aspects.   I. Title
QH445.2.W55   1994
174′.25—dc20
93-26352
CIP

Photoset by Parker Typesetting Service, Leicester
Printed in England by Clays Ltd, St Ives plc

To the three women in my life,
with love

CONTENTS

Preface   ix

1   In Search of the Holy Grail   1

2   The Messenger of Inheritance   16

3   The Anatomy of the Human Genome   40

4   From Microbes to Men   55

5   The Human Genome Project   72

6   Mosquitoes and Morals   97

7   New Genes for Old   134

8   The Moral Consequences of
    Molecular Biology   166

    Index   192

It has become a cliché of our times that the pace of scientific and technological change has accelerated to dizzying proportions. It has become a cliché because it is true. In the Western industrialized world, for example, no one's life has remained untouched by the ubiquitous computer chip. Yet, when I was studying physics at university, they were still teaching the electronics of valves before they introduced us to really exotic electronic components such as transistors: chips did not even rate a mention. And I am not that old.

In the entire world, there can hardly be anyone who has not been directly or indirectly affected by the discovery in the late 1930s of atomic fission. This seemingly obscure branch of applied quantum mechanics – and what could appear more obscure than such a topic? – led to the atomic bomb and, later, to a weapon of theoretically unlimited destructiveness, the hydrogen bomb. Even the poorest peasant tending a paddy field in the remotest province of China is involved: if he pays taxes at all, he must pay more to support the development and manufacture of his country's nuclear armaments; if he is too indigent to pay tax then there is less money available to lift him out of poverty, because it has been pre-empted by the cost of the nuclear arsenal.

Now, as the century of science draws to a close, there is another development which promises to touch all our lives in the closest and most intimate way. After decades of painstaking research, seemingly disparate paths into the sciences of molecular biology, chemistry, biology and genetics have converged. Suddenly the scientists realize that they are not just in a clearing in the jungle but at the peak of a mountain where all the surrounding landscape is clear to their view. They are confident now that they can tackle one of the biggest and most profound issues in their science: unravelling the message of human inheritance. Within a little more than a decade, they believe, they will have located all human genes and picked through the double helix of human DNA so that they can load the entire set of genetic instructions that specifies a human being into computer databases.

Human reproduction – the getting and the bearing of children – and the course of human lives – not least the impact upon them of diseases such as sickle-cell anaemia and cancer – will be profoundly affected by the new human genetics. To eat of this tree of knowledge may not, quite, fulfil the serpent's promise that 'your eyes shall be opened and ye shall be as gods', but with this knowledge will inescapably come power also. How is this power to be exercised and who is to wield it? That is the central theme of this book.

One of the frustrations of molecular biology is that, even more than nuclear physics, its practitioners talk in what appears to be a foreign language. The concepts they employ are not necessarily more subtle than those of nuclear physicists or cosmologists, yet they are expressed in terms virtually incomprehensible to the ordinary reader. The geneticist inhabits a city of Babel in which heterozygotes carry domin-ant and recessive alleles at their genetic loci. Also, astronomers deal with incomparably remote celestial objects, yet everyone who has looked above their head at night is familiar with the stars. Geneticists deal with the stuff that makes us what we are, yet few outside the profession of biological science have ever seen their own chromosomes.

There is an undeniable arrogance on the part of some researchers, a sort of patrician disdain for the mass of the plebs. Our research, they seem to be saying, whether the masses know it or not, whether they like it or not, is going to change their lives, and if they want to comment on it or try to influence how it will affect them, then they must adopt our vocabulary and master our technical argot. Only discourse couched in our terms will be deemed legitimate.

Consequently, some books about the new human genetics read as if they were unofficial textbooks, laden with the vocabulary of the trade. I have tried a different approach and have ended up writing as much about history as biology. I hope I have succeeded in avoiding the language of the technocrats, although some specialized terms are unavoidable. In addition, I have tried to avoid a heavily technical work in the philosophical sense – in part because I am not competent to do so, but also because I feel that it is important to keep clear sight of what the science and the technology are likely to achieve and thus what the moral problems are likely to be. What lies beyond the horizon is invisible and describing the landscape which can be seen is task enough. The past can illuminate the present, and because I believe the story of how democratic societies have dealt with human genetics in the past is relevant to the present situation, I have recounted this story in some detail.

Any discussion of the moral consequences of the Human Genome Project must be founded on an understanding of both the powers and limitations of the science – and that requires history of science as much as basic facts about biology. The first section of this book presents a brief account of how the science of molecular biology got to the stage where so ambitious an initiative as the Human Genome Project could be contemplated, and describes how the project itself came into being. After the science comes the application, and with it the new choices for individuals and society. These are described in Chapters 6 to 8. One aspect of the Human Genome Project has received comparatively little attention even though it is likely to be one of the most significant in terms of its effect on our lives. After a gene has been identified and isolated, one of the first steps is to genetically engineer it into bacteria or some other microorganism so that the microbe mass-produces the corresponding protein. Having previously rare human proteins available in large quantities will be a novel situation, and some of the possible consequences are described in Chapter 7.

Like the sword of justice, technology has a double edge. It can be used for ill or for good. The technologies that will follow, that have already followed, from the advances in human genetics can alleviate human suffering and misery. But we, the laity and not the professionals, must master the new knowledge and its technological application. The only way to do so is to transcend the geneticists' approach – to remember that human beings are more than just vehicles for transmitting genetic information from one generation to the next, and that human life is more than just the expression of some computer program written in the biochemical language of DNA

I should like to thank Dr Jeremy Cherfas and Dr Richard Nicholson for encouragement and advice. I am also especially indebted to my editor Eleanor Lawrence and, for her guidance and encouragement, to Susanne McDadd. No book is ever really finished and the errors which remain are, of course, my own.

T. W.

1

In Search of the Holy Grail

IT has been called the Holy Grail of modern biology. Costing more than £2 billion, it is the most ambitious scientific project since the Apollo programme to land a man on the moon. And it will take longer to accomplish than the lunar missions, for it will not be complete until the early years of the next century. Even before it is finished, according to its proponents, this project should open up new understanding of, and new treatments for, many of the ailments that afflict humanity. As a result of the Human Genome Project, there will be new hope of liberation from the shadows of cancer, heart disease, autoimmune diseases such as rheumatoid arthritis, and some psychiatric illnesses.

The objective of the Human Genome Project is simple to state, but audacious in scope: to map and analyse every single gene within the double helix of humanity's DNA. The project will reveal a new human anatomy – not the bones, muscles and sinews, but the complete genetic blueprint for a human being. Just as Vesalius's first anatomical atlas inaugurated a new era of human medicine, so – the proponents of the Human Genome Project claim – the new genetical anatomy will transform medicine and mitigate human suffering in the twenty-first century. But others see the future through a darker glass, and fear that the project may open the door to a world peopled by Frankenstein's monsters and disfigured by a new eugenics.

The genetic inheritance a baby receives from its parents at the moment of conception fixes much of its later development, determining characteristics as varied as whether it will have blue eyes or suffer from a life-threatening illness such as cystic fibrosis. The human genome is the compendium of all these inherited genetic instructions. Written out along the double helix of DNA are the chemical letters of the genetic text. It is an extremely long text, for the human genome contains more than 3 billion letters. On the printed page it would fill about 7,000 volumes the size of this one. Yet, within little more than a decade, the position of every letter and its relation to

I

its neighbours will have been tracked down, analysed and recorded.

Considering how many letters there are in the human genome, nature is an excellent proofreader. But sometimes there are mistakes. An error in a single 'word' – a gene – can give rise to the crippling disease of cystic fibrosis, the commonest genetic disorder among Caucasians. Errors in the genetic recipe for haemoglobin, the protein that gives blood its characteristic red colour and which carries oxygen from the lungs to the rest of the body, give rise to the most common single-gene disorder in the world: thalassaemia. A different error in the same gene – a mistake in just one letter among those 3 billion or so – is responsible for another of the most widespread inherited diseases, sickle-cell anaemia. More than 4,000 such single-gene defects are known to afflict humanity. The majority of them are fatal; the majority of the victims are children. The distress and suffering caused by genetic disease embrace not just those with the disorder but also their parents, who must live with the knowledge that they have passed on the defect as part of their child's genetic inheritance.

None of the single-gene disorders is a disease in the conventional sense, for which it would be possible to administer a curative drug: the defect is preprogrammed into every cell of the sufferer's body. But there is hope of progress. In 1986, American researchers identified the genetic defect underlying one type of muscular dystrophy. In 1989, a team of American and Canadian biologists announced that they had found the site of the gene which, when defective, gives rise to cystic fibrosis. Indeed, not only had they located the gene, they had analysed the sequence of letters within it and had identified the spelling mistake responsible for the disease. At the least, these scientific advances offer a way of screening parents who might be at risk of transmitting a single-gene defect to any children that they conceive. Foetuses can be tested while in the womb, and if found free of the genetic defect, the parents will be relieved of worry and stress, knowing that they will be delivered of a baby free from the disorder.

Then, in September 1991, came an astonishing advance: a four-year-old girl became the first patient ever to undergo a successful gene transplant in the treatment of a fatal genetic disorder. Intact copies of a gene were inserted into cells of her body to compensate for the nonfunctioning of the defective copies which she had inherited. The transplanted genes provided a new set of instructions for the affected cells which rectified her inborn defect.

In the mid-1980s, the idea gained currency within the scientific world that the techniques which were successfully deciphering disease-related genes could be applied to a larger project: if science can learn the genetic spelling of cystic fibrosis, why not attempt to find out how to spell 'human'?

Momentum quickly built up behind the Human Genome Project and its objective of 'sequencing' the entire genome – writing out all the letters in their correct order. It is the boldest and certainly the biggest endeavour ever undertaken in biology. It will make the twenty-first century the age of the gene. Although the Human Genome Project can be compared to the Apollo programme, it will transform human life and human history more profoundly than all the high-tech inventions of the space age. Its impact will far exceed the understanding and treatment of the single-gene defects already mentioned. There is clear evidence that other diseases, such as diabetes, heart disease, cancer and some psychiatric illnesses, although not inherited in a straightforward manner like sickle-cell anaemia, none the less have a strong genetic component.

But the ramifications of the project now under way in laboratories and research institutes around the world go far beyond a narrow focus on disease. Some of its proponents have made claims of great extravagance – that the Human Genome Project will bring us to understand, at the most fundamental level, what it is to be human. Yet there is a legitimate worry that such an emphasis on humanity's genetic constitution may distort our sense of values, and lead us to forget that human life is more than just the expression of a genetic program written in the chemistry of DNA.

If properly applied, the new knowledge generated by the Human Genome Project may free humanity from the terrible scourge of diverse diseases. But if the new knowledge is not used wisely, it also holds the threat of creating new forms of discrimination and new methods of oppression. No one will be untouched, for we have all been formed by the genetic inheritance we have received from our parents. Many characteristics, such as height and intelligence, result not from the action of genes alone, but from subtle interactions between genes and the environment. What would be the implications if humanity were to understand, with precision, the genetic constitution which, given the same environment, will predispose one person towards a higher intelligence than another individual whose genes

3

were differently shuffled? The Human Genome Project holds the promise that, ultimately, we may be able to alter our genetic inheritance if we so choose. And there is the central moral problem: how can we ensure that when we choose, we choose correctly? That such a potential is a promise and not a threat?

Once before in this century, the relentless curiosity of scientific researchers brought to light forces of nature, the mastery of which has shaped the destiny of nations and overshadowed all our lives. After the mushroom cloud rose over the New Mexico desert to usher in the atomic age in 1945, J. Robert Oppenheimer, the bomb's creator, remarked, 'We knew the world would not be the same.' Mankind's mastery of its own genes will not be demonstrated in the same dramatic fashion, but the work of the genetics laboratories will change the world as surely and perhaps even more profoundly than the physicists of Los Alamos. By creating a weapon of mass-destruction, Oppenheimer concluded that 'the physicists have known sin; and this is a knowledge which they cannot lose.' The choices before the geneticists are less clear cut, more subtle, and carry the risk of sliding unawares into sin while trying to do good.

While Alamogordo on 16 July 1945 serves as a convenient marker for the birth of the atomic age, the age of genetics will never have such a clearly defined parturition. But one can at least date its conception. In 1953 at the University of Cambridge in England, two brilliant young men, James Watson and Francis Crick, discovered that DNA, the molecular messenger of human inheritance – and of the inheritance of every living creature on the face of the earth – carried its information within the spirals of a double helix. Although our times have seen scientific discoveries of colossal import, ranging from Einstein's theory of relativity to the nuclear fission which fires the atomic bomb, several Nobel prize-winners have described Watson and Crick's achievement as the twentieth century's greatest single scientific discovery.

Nearly four decades later, after sharing the 1962 Nobel Prize for the discovery of the double helix, James Watson applied his genius to leading the United States' expedition into the genetic structure of mankind. The Human Genome Project is an international effort: there is not one but many projects, with researchers in different laboratories and in different countries pursuing the genes that most interest them. The US programme is far and away the best financed and co-ordinated. Somewhat unusually, it is being run jointly by the US National Institutes of Health (NIH), the body which funds medical research in the USA, and by the US Department of Energy. For the first three years, from 1989 to 1992, Watson was director of the lead agency, the National Center for Human Genome Research, which had been set up by the NIH.

Knowledge of the human genome, James Watson believes, will open up a pathway to a goal that touches all humanity: it will lead to a reduction in the human suffering caused by genetic disease. The co-discoverer of the double helix now works from a quiet office at the Cold Spring Harbor Laboratory on New York's Long Island, one of the world's leading genetics research institutes, where he has been director for a quarter of a century. The cruelty of genetic mistakes and the misery which they bring to human lives are leitmotivs in his conversation. Medicine in the twenty-first century will be dominated by genetics, according to Dr Watson, and the medical implications are driving the US genome project. Because there is a genetic component to many, more complicated illnesses, from late-onset diabetes through possibly to arthritis, genetic diagnosis need not be confined to the prenatal screening of foetuses at risk of severe genetic defects. A way is opened to preventive genetical medicine. If, from an analysis of their genes, someone were discovered to have a predisposition to diabetes, say, then they could modify their diet from an early age and so reduce the chances that the condition would actually develop.

Economists and industrialists are already looking at the commercial possibilities opened up by the genome project. According to one estimate, by the British Centre for the Exploitation of Science and Technology, products derived from genome research may account for drug sales of $60 billion or more by the year 2010, equal to half of the international pharmaceutical industry's sales in 1992. The Centre estimates that it will take some $30 billion of 'development' to convert the raw data from the project into marketable products. Enormous though this figure may seem, it represents only two years of the worldwide pharmaceutical industry's current R&D expenditure.

But there are dissenting voices. Even within biology, some professional scientists are sceptical. Some American researchers have actively opposed the rise of large centralized bodies such as the US National Center for Human Genome Research, arguing that due to its high profile it will attract money that otherwise would go to less fashionable but possibly more useful research projects. They point to the rise of 'Big Science' in other subjects – such as particle physics, dependent on huge atom-smashers maintained by armies of technicians and support scientists – and they see the problems of management that such enterprises inevitably create. Even the proponents of the genome project accept that when large-scale sequencing begins in earnest, it will inevitably have to be done on a quasi-industrial basis. The sceptics are concerned, too, about *dirigisme* from the centre: they fear that, instead of being able to pick their own topic for research – to pursue the ideas that they think will be scientifically interesting and productive – they may be assigned the task of sequencing a boring stretch of DNA.

In the very early days of the American project, Watson gave grounds for such concern, suggesting that the longest and most boring bit of the genome should be given to the Soviets to sequence. Such remarks have long since been withdrawn. More controversially still, he at one point appeared to suggest that the USA would keep its data confidential if other nations did not contribute equally to the project. The idea of the USA or any other nation having a monopoly of the human genetic blueprint was contentious to say the least. But as the international Human Genome Organisation (known inevitably as HUGO), which was founded in 1988, gradually started to play its role in co-ordinating international efforts, such threats receded and harmony was restored among the laboratories. However, the issue was given a new lease of life in 1991 by moves from another department within the NIH to patent every stretch of human DNA that it could, regardless of whether the scientists have any idea as to what function the gene serves. This has prompted other nations – notably Britain – to warn that they would follow suit and patent any DNA sequences that they have isolated. At the time of writing, the entire edifice of international scientific collaboration seems to be endangered by a squalid squabble over who will have commercial rights to human DNA. It is a tangled tale, not yet resolved, and will be explored in greater detail later in this book.

Disputes among scientists over the funding for research projects are nothing new; the squabble over patent protection for the fruits of the research raises issues of equity, but is at base a commercial disagreement. Others have put forward objections to the genome project that focus on deeper moral rather than monetary questions.

Watson concedes that 'As more and more things are revealed to have genetic components, we may see a feeling of losing control over our own destiny.' He stipulated at the beginning that 3 per cent of the money spent on the US Human Genome Project should be devoted to examining the social, ethical and legal implications of the work. It is now the biggest bioethics research programme in the US and, indeed, the world. And the amount of money being spent is increasing. 'We are spending 5 per cent of our money on worrying about the consequences, and that's going to go up,' Watson has said. 'As we learn more and more, civilization can't continue to ignore what we are, what the nature of life is, and the social implications of what we're learning about biology.' He believes strongly that if society is to cope with the consequences of this knowledge, people must learn and become better informed about genetics. The whole question of the nature and value of human life will inevitably be raised. 'It's not as simple as it was when you thought all your instructions came from the skies, and the rules for human behaviour with them. We don't have that consolation.' But Watson rejects pessimism. The new genetic technologies will not degrade our values: 'I don't believe that a lessening of respect for human life will occur.' And the advantages will outweight the potential abuses: 'Disease is never ennobling, and to the extent that the Human Genome Project can prevent disease, human life will be better.'

Most geneticists believe, like James Watson, that genetics must ultimately serve human ends. Moral problems arise in trying to define those ends and in ensuring that the means to attain them are ethically acceptable. Some very distinguished geneticists maintain that the new genetics does not raise any moral or ethical issues that are really new: what is being done in the laboratories and genetics clinics in the industrialized world, they say, is simply an extension of previous work, raising no new issues of principle. Although this is true, it misses the point. The investigations into human genetics now being pursued under the aegis of the Human Genome Project will present the old moral dilemmas in new guise. The scientists do not regard it as a trivial task to extend their present knowledge about human genetics; similarly, there is no reason to dismiss as trivial the process of extending our moral understanding of the consequences of this research – even if the basic moral principles remain unchanged. When President Kennedy decided in the 1960s that Americans

7

would go to the moon, he did not set aside 3 per cent of the budget to consider the ethical implications of doing so. It is also worth noting that, by and large, those who dismiss the ethical implications of the new genetics tend to be laboratory-based scientists remote from the clinical application of the technology. Watson is of course a distinguished exception. Those who have every day to go into hospital clinics and face parents at risk of having children with genetic defects are much more apprised of the human and moral impact of the new techniques. In the UK for example, Sir David Weatherall, one of the most distinguished researchers into hereditary disorders of the blood and a practising clinician, has taken a leading role in promoting discussion both within the medical profession and within government in considering the broader consequences of the new genetics.

One medical researcher who has been forced to confront the moral choices opened up by an earlier technology which he helped create is Professor Robert Edwards, who together with Patrick Steptoe pioneered the *in vitro* fertilization technique which led to the birth of the world's first 'test tube' baby. In a speech to the Centre for Social Ethics and Policy at the University of Manchester in 1987, Professor Edwards warned against looking to scientists for moral guidance about the consequences of their work:

Scientists are notoriously shy of ethics in relation to the general public. Many of them do not care to enter such debates even in their own field of work unless the social circumstances literally compel them into the ethical discussion. Most scientists have never been trained in ethics and they face considerable difficulties when faced with the formulation of their own ethical principles in relation to their subject.

The obverse of that coin is that few moral philosophers have been trained in science and so, in considering the moral implications of the new genetics they may, for example, devote too much attention to topics which are only remotely probable and which therefore should not occupy too much of our energies. Moreover, the real world is wider and more complicated, and human relationships are downright messier, than the abstractions contemplated in either laboratory or philosophy seminar.

Ultimately, much of the new knowledge generated by the Human Genome Project will be translated into products sold for profit in the open market-place. Will moral principles or commercial considerations guide the choice of what discoveries are exploited? Although morality and commerce need not conflict, the issue of patenting gene sequences has already demonstrated the tensions that exist. In the democracies, the elected representatives of the people will be the final arbiters of such disputes. But most legislatures already have full agendas and the people's representatives have many calls on their time. There is a need to spell out in advance which aspects of the new genetics are most likely to lead to difficulties and to suggest how society as a whole might avoid future trouble.

Scientists tend to assume that the knowledge they produce can be assimilated and utilized by the existing institutions of society. But it may not be so. Some of that knowledge may in itself be subversive of existing institutions. In that case, we can expect that that aspect of the science will either not be pursued or, if pursued, its results will be slow in the application. There is a clear example of this from the recent past: cigarette smoking. Nothing could be clearer than the demonstration of the connection between smoking and cancer and heart disease. This is no longer news, but when it was first put forward by the epidemiologist Sir Richard Doll in the 1950s, it was a novel finding. After the first scientific papers were published, the tobacco companies questioned the scientific evidence for decades, pursuing scepticism long after the scientific consensus had decided that the connection was unmistakable. Having failed to undermine the scientific evidence, they then employed arguments framed in the moral language of individual rights, alleging that it would be an incursion on personal freedom to curtail smoking. And so the companies, their advertising agencies and even, until recently, the press and TV which depend on advertising revenue, have all supported the continuing manufacture of a product that was known to be carcinogenic and which – had we known when smoking was first introduced what we know now – would never have been permitted on the grounds of danger to public health. Even governments have been slow in moving against the tobacco lobby: for fear of party political disadvantage, because the industry is skilful at finding the right places to lobby (lobbying is more powerful than the laws of science or the logic of moral argument in formulating the laws of the land), and because of the fears of unemployment among the workers in the industry (and hence more political unpopularity). Society's capacity for coping with

9

new knowledge is limited when that knowledge runs up against vested interests.

If the central moral issue is that of choice, of choosing correctly among the new options which the Human Genome Project will open up, then an essential first task is to identify properly what these options are likely to be. Are we really on the verge of the 'gene supermarket', for example, where prospective parents can go shopping for the traits they would like to see expressed in their children? Or is genetic technology never likely to be so far advanced? For the moment at least, the more lurid fears of wholesale Frankensteinian tinkering with humanity's genetic inheritance can be discounted. The genome project is first and foremost a research programme to discover new knowledge about human genetics and so the first issue to be faced is the handling of that knowledge: who has a right to own information about human DNA and who should have access to that information? As noted earlier, the patenting of human genes has already raised such questions from the point of view of national interest and commercial advantage. But for all the international acrimony which the patenting issue has raised, few national authorities are opposed to the principle of patenting human genes; the dispute is over the timing. The belief is that proper patent protection will give commercial pharmaceutical companies the confidence to invest their own money in developing genetic diagnostic kits and drugs.

Genetics diagnosis also produces new knowledge – in this case, not a generic understanding of human heredity, but specific information about an individual's genetic inheritance. Here too there are problems of right of access to such information. People with an abnormal form of a gene may be at risk of bearing children with the corresponding disease or may be at risk of developing disease later in life. The crucial point is that a genetic diagnosis will be done months, years, sometimes even decades before the first symptoms show themselves. That puts an all but intolerable burden of foreknowledge upon the prospective parents – if the diagnosis is done before a child is born – or upon an individual, if the examination is carried out on an adult at risk of disease later in life.

At first glance, it might appear as if there were no property so private, no information more personal and confidential, than a human being's own individual genetic blueprint. But a moment's consideration will show that this is not true. If I am carrying a potentially lethal genetic disease which will only show itself in a couple of decades, then my spouse surely has a right to know this now, for the sake of children as yet unborn. If I wish to take out a life assurance policy, will the company insist that I have the relevant genes analysed? And if I have the 'wrong' genes, will they be able legally to refuse to issue life cover?

The state too has an interest in the genetic health of its citizens, for it must underwrite the costs of medical care for those who suffer illness partly as a result of what is written in their genes. There is a risk that knowledge generated by the Human Genome Project could lead to the creation of a subcaste of genetic lepers who are refused jobs, insurance cover, even possibly the right to marry and have children. With such possibilities, the impact of the new genetics is not confined solely to the arena of what might be called 'personal morality': there will be consequences for the shape and organization of broader society. It is not putting it too extravagantly, for example, to suggest that a society in which genetic testing is widely available might prove to be incompatible with a system of private health insurance such as obtains in the USA, and can be compatible only with a 'socialized' medical system, such as the National Health Service of the UK.

Some people in the United States have already begun to think about these issues and in September 1991 California's state legislature voted for a genetic privacy bill, which would have outlawed genetic discrimination and imposed an eight-year moratorium on the use of genetic test results by health insurers. It is worth noting that California's response to scientific advance was not to change its pre-existing social and economic structures, by reforming the system of health insurance, but to try to stop the application of the science. The attempt was in any case defeated, when the bill was vetoed by the state's Governor Peter Wilson on 14 October 1991. The European Parliament is also looking at the issue of genetic privacy and suggesting that insurance companies should not be allowed access to the results of genetic tests. As matters stand now, if I should discover that I am carrying genes that would particularly predispose me to heart disease or some type of cancer, the law expects me to inform my insurance company. Not to do so would be to act in bad faith, and would render the insurance contract void.

If people do not have a right to genetic privacy, then do they have copyright to their own genetic information? It appears unlikely. In the United States, the courts have decided that people do not own the tissues of their own bodies. In 1976, one John Moore had his spleen removed as part of a treatment for a disease called 'hairy-cell leuk-aemia'. His doctors realized that the cells in the spleen produced a particular type of protein that could be useful in fighting cancer, so, after modifying the cells to get them to reproduce indefinitely in the laboratory, they patented the resulting 'immortal' cell line taken from Mr Moore. The Californian supreme court ruled in mid-July 1990 that Moore has no right to share any profits made on drugs derived from the cell line taken from his own body. The Moore case predated the human genome initiative and is not a clear-cut test of genetic copyright. But the genome project will bring these issues to the fore.

The moral dilemmas raised by testing for genetic disease are reasonably clear compared with some of the questions that might arise. Consider what might happen when it becomes possible to test, not just for the genetic signature of a disease, but for other, non-pathological traits. It is conceivable that within a few decades researchers may have worked out genetic profiles characteristic of enhanced intelligence, for example. Would we then see some parents conceiving children and selectively aborting until they obtained a foetus with the desired genetic profile? Repeated conception followed by abortion may not be necessary: it would be possible to opt for *in vitro* fertilization, screening the fertilized eggs to implant only the one with the most favourable genes.

A society in which such things occur, even if rarely, might seem deeply unattractive, with human life reduced to little more than a genetic specification and parents no longer wanting what is best for their child but calculating how to obtain the best child. Yet in some Western societies abortion is today available virtually on demand, so it is not easy to argue that the motive of seeking a foetus of higher intelligence would be morally more repellent than some of the reasons for which abortions are already performed. It is also difficult to see how such practices could effectively be forbidden. In some societies, where male children are preferred, parents have already made use of currently available technology to determine the sex of a foetus and selectively abort those which are female.

Because the subject is changing so rapidly, what is technically achievable in the field of human genetics now will be out of date tomorrow. The eminent moral philosopher Bernard Williams touched on this when he spoke on the ethics of the new genetics at a conference organized by the Ciba Foundation in Switzerland in 1989. Professor Williams said: 'It is a requirement on moral argument that it shouldn't simply stop at mere technical fact and say that the question does not yet arise.' He counselled against the view that the best way to diminish public fears about the direction in which genetics was leading was to emphasize what is impossible in practice today. Scientists, he believed, must look to the future as well, and philosophers could help them by suggesting how much further it was morally acceptable to go. There are issues not immediately linked to technical feasibility which must be explored in relation to the Human Genome Project. Could the elucidation of our genes lead to a moral vacuum in which human life was regarded as no more than the biological expression of a set of genetic instructions – a sort of biochemical computer program?

Once we know the program written into our genes by nature, the temptation to change a few bits here and there might well be irresistible. Less than half a century ago, Germany, one of the most cultured and civilized of European nations, embarked on a plan for genetic improvement with appalling results. It is commonplace to draw a complacent conclusion that Germany was at the time in the grip of a fascist and racist ideology, from which the world at large has recoiled in horror and which surely can never be repeated. Yet the record of the world's democracies in relation to genetics is not comforting. Eugenic ideas involving the forcible sterilization of the 'feeble-minded' were current in both Britain and the USA during the early years of this century. Before the First World War, Winston Churchill, then Home Secretary, tried to introduce a programme to improve the stock of the British race because, he believed, the feeble-minded were outbreeding the intelligent. In the USA, such programmes were carried into effect and thousands were sterilized. Within the past twenty years, India, the world's largest democracy, forcibly sterilized its citizens as part of its birth control programme.

These were all crude interferences with the rights of individual citizens to decide their own reproductive future – but what might be the results of employing the sophisticated technology of today's molecular biology to tinker with humanity's genetic structure? Those who have the skill to carry out genetic manipulation, and those who

13

pay them, may have access to great power in the future and they, or the institutions which they serve, may seek for themselves a monopoly of choice on how that power should be exercised. If, in a free society, the first maxim must be to distrust those having power, what constraints if any should be put upon the activities of those investigating human genetics?

In Britain, there is no national forum where such issues can be debated and a balance sought between scientists' freedom to work on what they perceive to be the most challenging and interesting problems in biology and society's ability to cope with the consequences of these researches. Many senior medical researchers would like to see a standing Royal Commission on Bioethics (rather like the existing Royal Commission on the Environment) to consider the various issues being generated by advances in modern biology – the Human Genome Project amongst them – but Conservative governments frown upon such innovations. One of the great charitable organizations, the Nuffield Foundation, has taken the initiative to set up an unofficial Council on Bioethics to keep science and society in harmony.

The role of the European Community will become increasingly important. In 1989, the European Parliament adopted the so-called 'Rothley report', which would severely limit work on the Human Genome Project. In practical terms this will have little effect – nothing happens as a result of the European Parliament's 'initiative legislation' unless and until it is adopted by the Commission. But the Parliament's action does represent the force of strong influences within the Community. The German Minister for Research has persuaded the EC Council of Ministers to set up an ethics committee to oversee the implications of the new biology – genetic engineering of plants and animals as well as the Human Genome Project. With its hideous experience with Hitlerian eugenics, Germany is of course particularly sensitive about issues involving human genetics. In 1993, European Justice Ministers through the Council of Europe were drawing up a 'Convention for the protection of the human person with regard to biomedical science'.

Other countries are more advanced than the UK in considering such issues. Australia, Denmark, France and Italy all have government-sponsored commissions investigating the ethical and legal implications of modern medicine and biology. In the United

States, many of these issues have been studied by the President's Commission for the Study of Ethical Problems in Medicine and Biomedical and Behavioral Research. Ironically, the commission did its work before the main lineaments of the new technology became clear and, as the science was progressing, President Reagan abolished the commission in the early 1980s. To some extent, the ethics research which is proceeding in parallel to the scientific aspects of the NIH's Human Genome Project will involve updating and revising some of the commission's conclusions.

Throughout this book, I shall treat the words 'moral' and 'ethical' as roughly synonymous. The term 'medical ethics' usually refers to the code of professional practice worked out for doctors, physicians and surgeons to regulate their dealings with their patients, but this is a huge subject in its own right and will be treated only tangentially in this book. Moral questions are much wider than those of professional practice in the clinic and consulting room, and it is not enough to consider the moral consequences of new scientific knowledge in isolation: one must also consider the social implications, how society (and not just the individuals within it) will cope. There is little tradition in the UK of thinking these issues through – what in the USA are called matters of public policy.

The options opened up by the Human Genome Project will be new, but the fundamental moral principles which will guide how we choose among them are not. Others have made analogous choices in the past, and the second part of this book examines what choices were made and how they were made. This emphasis on the past may seem strange in a book whose subject matter is the future. But this is not so. Knowledge of the past is an indispensable guide to the future. For, as Santayana warned, those who do not remember history are condemned to relive it.

3% put aside for researching ethics.

15

2

The Messenger of Inheritance

THE US National Institutes of Health (NIH) occupy a large collection of modern buildings sprawling over a green and leafy site in Bethesda, Maryland, just half a dozen stops on the metro from downtown Washington. It was there, on 14 September 1990, that the first successful attempt was made to treat disease by transplanting human genes into the body of the patient. A team of four medical researchers – W. French Anderson, Michael Blaese, Ken Culver and Steven Rosenberg – infused genetically engineered white blood cells into a four-year-old girl whose immune system was crippled by an inherited defect. Five months later, Dr Blaese told the 1991 annual meeting of the American Association for the Advancement of Science in Washington that the genetically engineered cells appeared to have had the intended effect of boosting the girl's resistance to disease. Within eight months of her first treatment, the little girl was going ice skating and taking dancing lessons – a remarkable transformation for someone whose life until then had been dogged by the fear that exposure to even the slightest infection could prove fatal.

The girl's body could not produce an essential protein – the enzyme adenosine deaminase (ADA). Without this enzyme her immune system was unable to develop or function properly. ADA deficiency is one cause of a condition known as severe combined immune deficiency (SCID) and accounts for between a fifth and a quarter of all cases of SCID. Those who suffer from SCID do not possess the natural armoury of defences against invading bacteria, viruses and fungi, an armoury that keeps the rest of us alive and healthy. They are fatally imperilled by infections that would in other people cause no more than a sniffle. Most children born with SCID die young – usually around the age of two – from overwhelming infection. In one well-known case, a boy in Houston, Texas spent his brief life inside a plastic bubble in an effort to isolate him from contact with harmful germs. With a worldwide incidence estimated to be between 1 in 100,000 and 1 in 500,000 births, SCID is mercifully rare, although knowledge of its rarity does little to lessen the tragedy for those families it afflicts.

SCID is not an infectious disease, nor is it the result of some external injury. It bears no relationship to Acquired Immune Deficiency Syndrome (AIDS) which, although it too results in the sufferer's immune system being crippled, is an infectious disease transmitted by a virus and which acts on the immune system by a completely different mechanism. Rather, SCID is part of the sufferer's being, as inborn as the colour of their eyes or the pattern of their fingerprints. There was a flaw in the genetic inheritance which the girl received from her parents: the instructions – the gene – for making ADA were faulty and the gene did not work properly. The white blood cells that Blaese transfused into her had been genetically modified to contain a correctly working normal copy of the ADA gene.

ADA is a protein – one of the classes of molecule essential for life. Many tens of thousands of different proteins (the exact number is unknown) go to make up the structure of the human body and to ensure that it functions properly. They include haemoglobin, which gives red blood cells their characteristic colour and transports oxygen from the lungs to the rest of the body, and collagen, which represents about one third of all the protein in the human body and is the major protein in tendons and cartilage. The enzymes – the essential biochemical catalysts – are proteins, as are the antibodies, which the immune system produces to combat infection, and many hormones, such as the insulin responsible for regulating the levels of sugar in the blood.

The instructions which a cell 'reads' in order to make a particular protein are embodied in a gene, which is sometimes said to 'code' for that protein. Genes act not just during the growth and development of an embryo; all cells are constantly reading the genetic recipes for proteins and then 'expressing' the genes by making the proteins. As I sit writing, my immune system is making antibodies to combat the slight cold that I am suffering from; my hair is growing, imperceptibly perhaps, but the cells are making fresh protein; without being conscious of the fact, I am digesting my lunch and this requires enzymes to aid the breakdown of the food; and the red cells in my bloodstream last only about 120 days, so my body is busy replenishing the supply and making copious amounts of haemoglobin in the process. All this and more is the effect of my body's cells reading instructions written in my genes and translating those instructions into the proteins that I require to live. The human genome is the compendium of all these genetic instructions and every cell in the human body contains its own private copy of the entire human genome. Each living organism, from bacteria to man, has its own characteristic genome.

Human beings inherit two complete copies of the genome – one from their mother and one from their father. In many cases the maternal and paternal copies are slightly different: one codes for blue eyes, say, and the other brown; one may specify blood group O, while the other specifies blood group A. For reasons that are still not entirely clear, some variants of a gene seem to dominate over others. Thus a child with mixed inheritance for eye colour will usually turn out to have brown eyes, although he or she will still carry the 'blue' gene and be able to pass it on to the next generation. The fact that everyone has two copies of the human genome in their body cells and the two copies are not identical to each other is, ultimately, the genetic basis of family resemblance. It is part of the reason why children look like each of their parents, but are not identical to either of them.

So to understand why the little girl in Bethesda could not make ADA, it is necessary to consider her parents. Her mother, unknowingly, had inherited a defective variant of the ADA gene from one of her own parents. That did not matter, as far as the mother's health was concerned, because this deficiency was compensated for by the 'normal' copy of the gene she had inherited from her other parent. The girl's father too had inherited one defective and one normal variant of the gene, and was perfectly healthy and unaware of this flaw in his genetic inheritance. When the girl inherited one copy of the ADA gene from her mother and one from her father, she could have received the normal gene from each, or a normal copy from one and the defective copy from the other. But as ill chance would have it, she inherited defective variants of the ADA gene from both her mother and father: and without the proper instructions, her body could not make this essential protein. The defect involves just one type of protein and just one type of gene. ADA deficiency is one of more than 4,000 different inherited 'single-gene disorders' which are known to afflict human beings.

There are many diseases like ADA deficiency which do not show themselves unless someone has inherited defective variants of a gene from both parents. The most common genetic diseases in the world are of this type: cystic fibrosis, sickle-cell anaemia, and the related group of blood disorders known as the thalassaemias. They are known as 'recessive' conditions because the presence of only one defective copy of a gene in a person's genome does not show itself as an illness.

A second group of genetic diseases show different patterns of incidence in men and women and are thus said to be 'sex-linked'. Women are symptomless carriers, whereas the diseases themselves show up in men. In this group are conditions such as haemophilia and the muscle-wasting disease Duchenne muscular dystrophy. Because of the way in which sex is determined genetically (see Chapter 3), there are a few genes for which men have only a single copy, whereas women have the usual two. If one of these genes is defective it does not manifest itself in a woman because its effects are masked by the other, intact copy, but gives rise to disease in a son who inherits the defective variant, even though it is recessive.

A third category of single-gene defects comprises rare 'dominant' conditions, where the inheritance of even one damaged copy of a particular gene is enough to cause illness: the fatal neurological disease Huntington's chorea, and a few, very rare, instances of early-onset Alzheimer's disease are examples.

Not all heritable disorders need be the result of a defect in just one gene; several genes may act in concert to cause disease. Some people, for example, appear to have a predisposition to heart disease – they may be fit and take lots of exercise but none the less succumb to coronary disease at an unusually early age, as do several of their close relatives. Such predisposition to heart disease is an example of a polygenic condition. Such things are naturally more difficult to study than the case of single-gene disorders and so most research has focused initially on single-gene defects, in the hope not only of alleviating such disorders but also of using the knowledge gained to deal with the more complex polygenic diseases.

ADA deficiency can be tackled by bone-marrow transplants, as the cells mainly affected – the lymphocytes – are produced by the bone marrow. But transplants are only possible if the sufferer has a compatible sibling who does not suffer from the disease to act as donor. The donated normal bone-marrow cells contain the correct instructions for making ADA and, once transplanted into the patient's bone marrow, start making the required protein. It is also possible to give patients the ADA protein itself, but injections do not get it immediately to the right place in the right concentrations; they are a poor substitute for the subtle mechanisms controlling and directing production of ADA under normal circumstances, and the beneficial effects of such treatment tend to wear off after a year or so.

So it was that in the autumn of 1990, Blaese and his colleagues achieved something never before attained in human medicine: they successfully rewrote the defective genetic recipe for the ADA protein in the cells of the girl's body.

Genetic instructions are written out chemically in the composition of DNA, the molecular messenger of inheritance. In human cells the genes are kept locked up in the centre, or nucleus, of each cell and it is from this location that DNA (deoxyribonucleic acid) takes on its appellation of a 'nucleic acid'. DNA and proteins are the two most important molecules of life and there is a symmetry between DNA, in which the genetic instructions are written, and the gene products, the proteins, which are produced when the cells act upon or express the genes. In a rough and ready sense this is a one-to-one relationship: one gene, one protein. ADA protein was created for the girl by altering her DNA to give her an additional intact ADA gene.

DNA is an invisibly fine gossamer thread and no scalpel can be honed sharp enough, no surgeon's hand can be steady enough, to cut DNA at precisely the right place, excise a damaged piece and stitch in a replacement gene. Yet despite this, gene splicing has been commonplace in biological laboratories for two decades now. It was first done in bacteria, then yeast, then plants and animals, until in 1990 it was judged sufficiently safe to be used in humans. The long road to such technical mastery over the molecule of inheritance – starting with the study of organisms which appear utterly remote from human beings, but leading ultimately to a little girl lying in a hospital bed in Bethesda – will be described in the next couple of chapters.

To correct the ADA deficiency the gene therapists at the NIH first took white blood cells, known as T lymphocytes (T cells), from the girl. These cells form part of the immune system's response to infection. Because it is not possible to insert a gene directly into a T cell, the medical team first inserted a normal copy of the ADA gene into a special virus and then used the genetically engineered virus to infect T cells extracted from the girl and cultured in the laboratory.

Although it may seem paradoxical to treat a disorder of someone's immune system by deliberately creating a viral infection, the use of viruses to transfer DNA in this way is a well-established technique and the procedure worked. The virus had been genetically modified itself to insure that it could not be harmful, before the normal ADA gene was inserted into it. The genetically engineered virus did its work and inserted the correct ADA gene into the cells without doing any other damage. The genetically modified cells were cultured in the laboratory, and allowed to divide to produce a large number. The gene-corrected cells were then transfused back into the little girl's body. It was remarkably like a simple blood transfusion, lasting about thirty minutes, throughout which the girl was sitting up, wide awake, chatting to her doctors and parents. The transfused cells remained active for up to forty days after transfusion, according to the initial account of the operation. Most important, the corrected cells began producing the ADA that was previously lacking.

Although the operation is a masterpiece of modern medical technology, it is doubtful if it will produce a permanent cure. For all the cleverness of the genetic surgeons, there are still problems, and the little girl's immune system has not been restored to perfect function. The transplanted T cells are not immortal: eventually they will be broken down and will be replaced by new ones – ADA-deficient ones – being made continuously by the 'stem' cells in her bone marrow. The gene transplant was therefore more like a highly sophisticated way of delivering a drug – the ADA protein – to the right place, at the right time, and in roughly the right amount. The genetic defect in the girl's body cannot be permanently rectified unless transplanted genetically corrected cells can oust and completely replace the defective bone marrow.

At the beginning of this experiment no one thought that such an outcome might be possible, and so it was expected that those who suffered from the condition would have to undergo transfusions of their own, genetically corrected, T cells at intervals throughout their lives. The girl still has to receive weekly injections of ADA directly and must undergo infusions of the gene-corrected cells every three to five months. The stem cells in the bone marrow are effectively immortal, and the only way to effect a permanent cure would be to reprogramme these cells with the correct recipe for ADA. In September 1990, there appeared to be no way of doing this, but other researchers have subsequently developed highly sophisticated techniques for identifying and isolating stem cells. The genetic surgeons are working out methods to capitalize on these new techniques and are seeking approval to apply them to human genetic disease.

Rosenberg and French Anderson followed up the first ADA transplant with a second in January 1991 on a nine-year-old girl and, a year later, by applying a similar type of gene therapy to the treatment of cancer. Before they started to treat the girl in 1990, the team of researchers had previously transplanted genes into volunteers, not in an attempt to cure disease, but as part of a long and detailed programme to ensure that the techniques would be safe and would not have unintended side-effects. Present-day gene transplantation techniques cannot guarantee that the gene will be inserted into exactly the right place in the patient's DNA; there is a chance that it might be inserted in such a way as to disrupt the functioning of an essential gene, or even inadvertently to trigger latent cancer-causing genes. However, as gene therapy is currently being considered only for those with disease so grave that they already face the prospect of imminent death, a theoretical risk of contracting cancer at some unspecified future date is not likely to be considered an impediment to accepting the treatment.

The ethical issues raised by the girl's gene transplant had been exhaustively examined, with the conclusion that there was no moral difference between this kind of gene transplantation and transplanting tissues or organs, such as kidneys or corneas, which society has long accepted. For the transplanted genes were put only into 'somatic' or body cells and thus affected only the little girl herself: her 'germ' cells, the ones which give rise to her eggs and thus to any children she might have, would not have been affected by the operation and so the genetic constitution of future generations ought to be unaffected. None the less, Dr Blaese and his colleagues were careful to consult as widely as possible on the propriety of what they intended to do and to secure approval for the procedures from ethics committees of the NIH. For the girl's treatment in September 1990 was not the first attempt to cure an inherited defect by gene therapy in humans: that had been carried out nearly a decade earlier, by Martin Cline of the University of California at Los Angeles. But Professor Cline acted before he had secured all the required approvals for his work from the authorities. As it happened, his gene transplants, while they did not harm the patients, did not help them. Because he had extended his experimental procedures slightly without seeking approval for the extra development, his research was deprived of funds from the US public purse. These issues will be discussed in later chapters.

The ability to rewrite the genetic instructions inside a human being is the culmination of developments in molecular biology dating from the late 1930s. The first application to humans of this new science came just after the Second World War when, in 1949, American researchers showed that people who suffered from the crippling disease sickle-cell anaemia had an abnormal form of the protein haemoglobin in their red blood cells. This was the first human 'molecular' disease to be recognized as such.

Until the late 1970s, human proteins were more accessible and tractable objects of study than human DNA, and any research on the molecular structure of genes themselves usually had to be done by examining the genes of bacteria, viruses or other simple organisms such as yeasts. For this reason, human molecular biology for much of the post-war period proceeded by way of protein analysis: the gene products were studied rather than the genes themselves. Pioneering gene therapy such as that carried out at the NIH could not even have been contemplated if researchers had not, through years of painstaking work, identified ADA as a vital enzyme in the functioning of the immune system and recognized that its absence was the prime cause of some cases of SCID. From the chemical composition of the ADA protein, researchers were able to make a pretty good guess as to what the gene carrying the instructions to make ADA would look like. Techniques which became available in the 1970s then made it possible to identify and look at the ADA gene directly – and thereby enabled Blaese and his colleagues eventually to genetically engineer corrected versions of the gene into the girl's white blood cells.

Since the early 1980s, however, it has been possible for geneticists to work in the other direction. They can now sometimes pinpoint a human genetic defect and then work out which protein in the body

23

must be going wrong on the basis solely of the genetic information. Researchers discovered that they had a powerful new diagnostic toolkit with which to tackle the problem of human genetic disease. And as the new techniques for direct analysis of DNA revealed the underlying causes of several important diseases, scientists realized that for the first time they had a set of tools so powerful that they could think of analysing all human genes, not just those implicated in disease.

The first triumph of this 'reverse genetics' came in 1986, when Tony Monaco at the Children's Hospital in Boston and Louis Kunkel at Harvard identified the genetic damage responsible for Duchenne muscular dystrophy (DMD). This wasting disease of the muscles is one of the sex-linked genetic disorders and so affects boys almost exclusively, afflicting 1 out of about every 3,500 male births. The infants seem normal for their first few years, but between the ages of three and seven the wasting disease sets in and the boys develop a strange waddling walk. Progressive weakening of their muscles confines the boys to a wheelchair by the time they are about ten or twelve, and many of the victims die in their early twenties. It was clear to medical researchers that these deficiencies stemmed from a failure to make a protein which the sufferer's muscles needed for their normal functioning. But there are thousands of proteins in normal muscles, and biochemists had had no success in singling out the one which was deficient in boys with DMD.

However, in 1981 Kay Davies and J. M. Murray, at St Mary's Hospital Medical School in London, began looking not for proteins but for genetic markers: distinctive and recognizable stretches of sufferers' DNA which might be associated with DMD. The success of their work sparked international efforts to pinpoint the exact location of the genetic damage within human DNA. These efforts bore fruit five years later when Kunkel and Monaco located the gene which, when damaged, gives rise to DMD. From the genetic specification, they identified a hitherto unknown protein, which they called dystrophin. It was the first time that a human protein involved in disease had been identified from analysis of its gene.

Dystrophin is estimated to correspond to about 20 parts per million (0.002 per cent) of the protein content in a muscle cell, which explains why no one had implicated it in muscular dystrophy. Sufferers from DMD appear to have no dystrophin in their muscles, but there is a milder form of the disease, called Becker muscular dystrophy, whose sufferers appear to have the protein in a damaged form. Now scientists have identified the protein responsible for DMD, there is at least a possibility that a way may be found of making good the loss – in a manner analogous to the way in which the girl in Bethesda has been treated for ADA deficiency – and so prevent the suffering and early death that results from DMD.

Three years after the isolation of the muscular dystrophy gene, the success of reverse genetics was repeated with the most common single-gene defect to afflict northern Europeans: cystic fibrosis (CF). Somewhere between 1 in 20 and 1 in 25 of all Caucasians carries a defective cystic fibrosis gene inherited either from their mother or from their father. But as with the parents of the girl with ADA deficiency, the health of someone carrying just one defective copy of the gene is not impaired, and they will have no inkling that they carry only one correct copy of the CF gene. Problems arise only when a man and a woman who each carry one defective CF copy marry and have children. (The chances of such a marriage are about [1 in 25] × [1 in 25], that is 1 in 625.) As CF is a recessive genetic disease (like ADA) there is a one-in-four chance (for each child) that a child born to two such CF carriers will receive a defective gene from each parent and suffer from the disease. The incidence of CF is roughly 1 in 2,500 births – about what would be expected from a one-in-four chance within 1 out of every 625 marriages.

Because it is the most common single-gene defect to afflict Caucasians, the symptoms of CF are comparatively well known. The lungs of affected children clog with thick, sticky mucus, which has to be cleared by intensive and regular physiotherapy. The pancreas, which produces enzymes essential for the stomach to digest food, is affected also and many sufferers experience digestive disorders caused by their bodies' inability to absorb fat and protein. A baby born with CF may appear normal at first, but eventually the symptoms begin to appear: persistent cough, wheezing, bouts of pneumonia, poor weight gain, and bulky, foul-smelling excreta. Sufferers also produce abnormal sweat and, nowadays, testing the sweat of infants suspected of having the condition provides a standard diagnostic test. In the past it was possible for years to pass before a definitive diagnosis was made, for some of the symptoms could result from common ailments of childhood. In the intervening years, other

25

children with CF might have been born into the same family. The mucus in the lungs acts as an ideal breeding ground for bacteria, so sufferers are liable to recurrent infections. In the days before antibiotics such infections were frequently fatal in childhood; even with the advantages of modern intensive medicine, the average life expectancy of those who suffer from CF is about twenty-seven years, although a baby born today with the disease might expect to survive to about the age of forty.

Until the late 1980s, there was no way of identifying those parents potentially at risk of producing CF progeny. Many people remained unaware that they were at risk of transmitting CF until one of their children was born with the disease. And the genetic lottery is blind to the concerns of humanity: it does not follow that because a couple have had one CF-affected child that their next three will be free of the condition. It is possible for one family to have two affected children in succession whereas another couple, who are also both carriers of a defective gene, might not give birth to an affected child. One of the urgent priorities in CF research was to develop a prenatal test so that a couple who already had one child with CF might be assured that a second pregnancy would not bring to term another child with the condition.

Unlike the case of ADA deficiency, years of research had failed to discover the single protein whose deficiency was responsible for CF. So, researchers turned to hunt for the signature of the disease within the DNA itself. They set out to look for genetic markers which were present within the DNA of children suffering from the condition but were absent from those who neither suffered from CF nor carried one defective copy of the gene.

The work lasted more than a decade and involved teams from Europe and north America. As with all quests there was not only the excitement of the chase, but also the frustration of sometimes finding that paths which had looked so promising turned out to be dead ends. The search for the gene developed into a hectic transatlantic race between researchers at St Mary's Hospital Medical School in London, headed by Professor Bob Williamson, a group at the University of Utah, headed by Ray White, and a US–Canadian team, led by Dr Francis Collins of the University of Michigan and Dr Lap-Chee Tsui in Toronto. Also involved was an American commercial company, Collaborative Research, which aimed to make money out of work on the human genome.

By 1984, the three groups had all located the approximate position of the gene. In doing so, they had detected characteristic patterns in the DNA of those families known to carry the CF mutation. They could thus already offer a genetic diagnosis of an unborn foetus to see if it carried the mutation in both maternal and paternal genes and so would inherit the disease. But this depended on isolating the pattern from individuals already known to be carrying the CF variant: without the gene itself, there was no way of conducting such a diagnosis in families which did not already have a case of CF. Then, in April 1987, Professor Williamson announced that he had discovered genetic markers which, he believed, pinpointed the gene's exact location. Alas, it was a false trail. Ray White had by this time dropped out of the race, and the search was to continue for a further two years until, in 1989, the US–Canadian group announced victory. They isolated the gene and started to decipher the instructions that it contained.

The gene implicated in CF contains the specification for a protein that forms part of the outer membrane of cells. This protein is believed to have several functions, at least one of which is involved in controlling the rate at which chloride ions flow into and out of the cells. Defects in the protein disrupt normal flow, giving rise both to the mucus within the lungs and to the excessively salty sweat produced by CF sufferers. Pinpointing the gene was a triumph of reverse genetics; there was little possibility of being able to identify the affected protein in any other way, because it forms part of the complex collection of proteins and other molecules making up cell membranes. But once they had identified the gene, by using their knowledge of the genetic code (see later in this chapter), the researchers could read off the composition of the protein and assess what effect the genetic defect might be having on it.

The success of the hunts for the DMD and CF genes is an astonishing demonstration of the power of modern molecular genetics which has been quietly growing up in laboratories and research centres around the world. They give a hint of the immense harvest that may be expected from this discipline.

More than 150 different genetic lesions have now been identified in CF sufferers. In many cases, the consequences are the same: the cell-membrane protein is deficient and unable to do its job properly, leading to the full-blown symptoms. Some mutations, however, lead to milder forms of the disease. Because of this variety of mutations in the gene (the commonest is present in about 78 per cent of British cases), direct gene testing of the population at large would miss some carriers of a defective CF gene and thus could fail to pick out couples who have no previous family history of CF but who might be at risk of having a CF baby.

Although the CF gene was found by American scientists, the first tentative steps towards screening the population for carriers of the CF mutation have been taken in Britain – the US has lagged quite considerably in the application of this new finding in the clinic. Screening people for possible genetic defects is a controversial and thorny issue. We all carry potentially lethal genetic defects: one estimate suggests that out of the 4,000 or so single-gene defects known, each individual probably carries about four or five mutations in their genetic constitution that would pass on a fatal defect to their children if, by some mischance, their partner carried the same defect. The issue of mass screening and what is to be done with the information gathered thereafter will be dealt with in Chapter 6.

The ability to analyse genes directly marks the start of one of the most exciting developments in modern science – what in 1980 the editor of the *American Journal of Human Genetics* called 'the new genetics'. These new techniques have brought hope that the suffering caused by genetic disease might be ameliorated. Their demonstrable power has quickened the scientific momentum behind the grand challenge of pinpointing not just disease-related mutations but identifying and deciphering all the genes in human DNA: the Human Genome Project. James Watson, co-discoverer of the double helix of DNA, described the rationale for the enterprise in a speech given in November 1990: 'Over the years, I have come to see DNA as the common thread that runs through all of us on this planet Earth. The Human Genome Project is not about one gene or another, or one disease or another. It is about the thread that binds us all.'

DNA is the thread that binds not only humanity, but all living organisms. For many years therefore, researchers chose simpler organisms than human beings in which to study DNA and its workings, secure in the knowledge that, however remote their studies might seem from the human condition, their research would one day illuminate human genetics because the general structure of the molecule of inheritance is held in common by all the living world. Yet it was only in the mid-1940s that DNA was identified as the carrier of inheritance. Before then scientists were convinced that proteins and not DNA carried the genetic information. Proteins were known to be biologically important, complicated molecules, whereas DNA was thought to be both a short molecule and a simple one chemically. Biologists also knew that proteins derived their biological activity from their three-dimensional shape. They naturally assumed that the information necessary to encode genes was somehow sculpted into the shape of special protein molecules which could act as if they were templates for their own replication. It was rather as if hereditary information was like a jigsaw, where the shape of the piece deter-mines where it fits, where only the shape matters and there is no analogue to the information contained in the picture that the whole jigsaw carries.

Both assumptions – that genes were carried by proteins and that the information content was conveyed by geometrical shape – turned out to be serious errors. In 1944, the American microbiologist Oswald Avery and his colleagues Colin MacLeod and Maclyn McCarty working at the Rockefeller Institute in New York showed that the message of inheritance was carried by nucleic acids and not by protein. It was almost as momentous a discovery, although much less publicized, as that of Watson and Crick. Watson is one of the authors of a standard textbook on the subject, *Molecular Biology of the Gene*, which describes the discovery simply as 'Avery's bombshell'. Over the next few years, more evidence accumulated pointing in the direction of DNA as the molecule of inheritance. The advent of electron microscopy after the Second World War also showed directly that DNA was a very long, thin molecule (although even an electron microscope is not powerful enough to display the double helix directly). In 1952, Alfred Hershey and Martha Chase, at the Cold Spring Harbor Laboratory on Long Island, New York, con-firmed Avery's earlier result. The obvious next step was to find out how DNA transmitted genetic information and the key to that was to unravel its detailed structure.

The story has been told many times of how the brash young American microbiologist James Watson and the English physics graduate Francis Crick discovered in 1953 that DNA has a double helical structure. It will not be repeated here, except to stress the extreme importance of the discovery. Watson was only twenty-five years old at the time; Crick, the older man whose career had been interrupted by the war, had not even secured his PhD. Together with Maurice Wilkins, who had been working on the problem at King's College London, they shared the 1962 Nobel Prize for what other Nobel Prize-winners have variously characterized as the greatest scientific discovery of the twentieth century, and one of the greatest discoveries ever made.

Yet although the DNA double helix has become almost an icon of our times, the shape was not in itself the most important aspect of the discovery. The two winding strands which make up the backbones of the double helix are pretty uniform and regular in their chemical composition. From the point of view of genetics, the most interesting part of DNA is not the spirals themselves but the cross-links that keep the two strands of the double helix together. These links wind up the inside of the double helix, like the steps on a spiral staircase. They are formed out of four different chemicals – the bases adenine (A), guanine (G), cytosine (C), and thymine (T). A, G, C and T are the genetic alphabet and it is the sequence of these letters and not the shape of the molecule which is important in determining our genetic inheritance: the picture on the jigsaw was what mattered after all.

Each base pairs up with a counterpart on the other strand of the double helix. The crucial point, which lies at the heart of Watson and Crick's original discovery, is that the way in which the bases pair up to make the steps of the spiral staircase is fixed chemically. If one strand of the double helix has the base A, then the other strand will have T in the corresponding place, so that the two always pair up. Similarly, C always pairs with G. The bases have different shapes and bind to each other differently so will not fit together in any other combinations. Thus, whatever the sequence of bases on one strand, the other strand must have the complementary sequence given by the pairing rules. The bonds between A and T and between C and G are comparatively weak, but there are millions of them running up the inside of a DNA molecule, and they act rather like the teeth of a zip. Individually they are weak, but when all are linked together at once, the bond is very strong indeed.

The pairing of A with T and C with G made sense of an earlier discovery in the chemistry of DNA. In 1950, Erwin Chargaff had analysed DNA from different organisms and showed that no matter where the DNA came from, the base A was present in roughly the same proportion as T; and similarly for C and G, whereas there was no simple relation between the quantities of C and T or A and G. Watson and Crick used Chargaff's rules about the relative amounts of the four bases to help them deduce the double helical structure.

Once Watson and Crick understood that DNA consisted of a sequence of base pairs which bridged the two backbones of the helix, they also had the key to understanding how genetic messages can be passed on from a parent cell to its progeny. Cells reproduce simply by dividing in two. After an egg has been fertilized by a sperm in the course of sexual reproduction, the result is a single cell, the fertilized ovum, which has its own unique complement of genetic instructions – its own genome. The egg then sets about growing: the single cell starts to divide to form two cells, which each divide to form four, which each divide to form eight, which each divide to form sixteen, which each divide to form thirty-two, and so on. The end result, in an adult human being, is about 10 million million cells. With the exception only of the sex cells – sperm or eggs – which will pass on instructions to the next generation, each of these cells contains a faithful and astonishingly exact copy of the genome that was present in the very first cell – the fertilized egg. The uniquely specific pairing of the bases across the strands of the double helix, as Watson and Crick realized, holds the key to making such faithful reproduction possible.

As each cell prepares to divide, it unzips the DNA double helix and uses each strand as a template to guide the formation of a new companion strand. In this way, the cell can make two copies of one DNA double helix, each containing one old strand and one new strand. When the cell divides, one copy passes to each of its daughter cells. The cell can be sure that the copies are faithful to the original, because where one original strand had A, then the cell must provide T; in the complementary original strand there will have been a T at the same place, so the cell must provide an A. Mistakes in replicating the DNA like this will be rare, because if the cell inadvertently fixed a G to an A, the new double helix would not zip together properly. Each cell has a panoply of proofreading mechanisms to ensure that mistakes are detected and repaired and the actual process of replication is of course a good deal more complicated than the outline sketched here.

As Francis Crick noted in his autobiography, *What Mad Pursuit*, DNA 'is indeed a remarkable molecule. Modern man is perhaps 50,000 years old, civilisation has existed for scarcely 10,000 years, and the United States for only just over 200 years, but DNA has been around for at least several billion years. All that time the double helix has been there, and active, and yet we are the first creatures on Earth to become aware of its existence.' There is an elegance and simplicity about the structure of DNA which is profoundly surprising when one considers the dazzling complexity and variety of life on earth. Yet the twist of the double helix provides an underlying unity for all life and the four letters of the genetic alphabet are all that Nature needs to spell out not just human beings, but every living thing.

A few viruses form an exception to this all-but-universal scheme, for their genetic code is written out not in DNA but in the closely related chemical, RNA (ribonucleic acid). Among the RNA viruses are some, such as tobacco mosaic virus, which afflict plants, but this class also includes agents of human disease such as polio, influenza and the AIDS virus. Many researchers, most notably the British-born Leslie Orgel now working in the USA, think that before the present variety of the living world developed there may have been a shadowy 'RNA world' in which all life was specified by instructions written on RNA rather than DNA. After the Earth condensed out of the swirl of dust and gases surrounding the Sun some 4.6 billion years ago, many of the chemicals essential to life, such as sugars and the bases, could have assembled themselves simply by the processes of ordinary chemistry on the surface of the primitive, lifeless Earth. Life began comparatively early: palaeobiologists have found evidence of fossil cells in rocks that are 3.5 billion years old.

Unfortunately, fossilization preserves only the outer structure of the cell walls and conveys no information about the molecular biology of the cell, nor do these fossils provide any clues about the life forms that preceded the first microorganisms to have tough outer cell walls. But there are, so to speak, 'molecular fossils' preserved within the structure of the DNA and RNA of present-day organisms which can yield some information on how life most probably started. Until the early 1980s, the problem of life's origin appeared to be insoluble: for in today's world, only nucleic acids contain genetic information, but they need proteins which act as catalysts to help them replicate.

Which came first, the protein or the nucleic acid?

Then in 1981, came the discovery that, under certain circumstances, some types of RNA could act as enzymes. This led to the suggestion that the very first living molecule might have been a stretch of RNA, created by an accident of chemistry, which catalysed its own replication without the help of a protein. Since the initial discovery, many more examples of RNA behaving like enzymes have been uncovered but, because it is chemically slightly different, DNA does not exhibit the same properties.

There are other arguments relating to the structure and present-day function of the two molecules which support the idea that some fragment of RNA was the first molecule capable not only of replicating itself but also of making errors in doing so and, therefore, of evolving. Eventually, the RNAs would have moved on from being their own crude catalysts and enzymes to manufacturing the more efficient proteins. But RNA itself is not as suitable as DNA for supporting the development of complexity: the molecule is not as chemically stable as DNA and is vulnerable to spontaneous breakdown. No organism as complex or as fast growing as a bacterium could evolve with an RNA genome, because of this chemical instability. The RNA world must have been over within about a billion years or so, although there is the possibility that some RNA viruses may be very ancient. Apart, therefore, from a few molecular fossils left over from that living world which preceded our own, for the past 3 billion years or so, DNA has been the molecule of inheritance for all living organisms.

While the elucidation of DNA structure solved the puzzle of how hereditary information was passed down the generations, it initially cast little light on the other important problem in molecular biology: how genes are 'expressed' to manufacture proteins. Throughout the lifetime of an organism, the biochemical machinery of its cells reads the genetic recipe and converts the instructions encoded in the sequence of As, Ts, Cs and Gs into the proteins which the cells need for their continued existence.

Proteins are constructed out of standard subunits, known as amino acids. Although only 20 different amino acids occur naturally, this is enough to give rise to the enormous diversity of different proteins. The analogy with the alphabet crops up here again: although there are only 26 letters in the English alphabet, that is sufficient to make many thousands of words which can make an almost unlimited variety of meaningful sentences. Proteins are like words built up from the amino acid letters, and a human being represents a vocabulary greater than Shakespeare's, for we consist of more than 50,000 different 'words'. Moreover, proteins are free from the preference for short words which constrains natural human languages: even German's capacity for concatenating smaller words into still larger ones is insignificant compared to the language of proteins, some of which may contain several thousand amino acid 'letters'. From dystrophin to haemoglobin, from ADA to the CF cell-membrane protein, all proteins are built up from the same set of 20 ingredients.

The function of a gene is to specify the order in which the different amino acids are strung together in a chain in order to construct a specific protein. Once made, the chain spontaneously folds itself up into a very precise shape – a shape determined by the order of the amino acids. Only in this configuration can the protein carry out its functions in the body. There is thus a profound correspondence between the composition of a gene and of the protein that the gene encodes: genes employ the language of bases; proteins, the language of amino acids; and living cells can translate one into the other. It took a decade and a half from the discovery of the double helix before scientists had entirely solved the problem of how the sequence of bases in DNA is converted into a sequence of amino acids.

At first, the old habit of thinking that the shape of the molecule contained the information persisted. In 1954, the physicist George Gamow suggested that the amino acids somehow fitted directly into 'holes' along the helix at specific places determined by the sequence of bases and that once so ordered, the amino acids would then join up to form the finished protein. But that was soon found to be impossible. The following year, Crick proposed that there had to be an intermediary between the sequence of bases along the gene and the amino acid sequence along the protein specified by the gene: some sort of 'adaptor' molecules which brought the appropriate amino acid to the correct place on the DNA strand. Crick's idea was the first step in understanding the complexities of protein synthesis.

The cells of the human body, in common with almost all living creatures other than bacteria, keep their DNA in an inner compartment called the nucleus. To make a protein, a cell will copy the relevant gene into an intermediary chemical, a molecular go-between, made out of RNA. It makes a single strand of this 'messenger RNA' which contains a sequence of bases exactly mirroring the gene's DNA sequence. The RNA then carries its message out of the nucleus so that the cell's biochemical machinery can read it and start to fabricate the protein. At this point Crick's 'adaptor' molecules come into play: they 'read' the sequence written in the strand of messenger RNA and string together the appropriate amino acids to form a growing chain, with the amino acids in the order specified by the messenger RNA. But how is the language of the genes constructed to give the message that the adaptor molecules read?

It turns out that the genetic code is couched in a series of three-letter words, with each word or 'codon' corresponding to one amino acid. There are no commas, spaces or punctuation marks between words along the RNA strand, except for the full stop which tells the cell's biochemical machinery that it has come to the end of a gene and that the amino acid chain should now be terminated. There is considerable redundancy in the system: since each of the four letters A, C, T and G can be used equally in a three-letter codon, there are sixty-four different codons ($4 \times 4 \times 4$). But there are only twenty amino acids. Several different words, or codons, can refer to the same amino acid and there are three different stop codons. The genetic code translating codons into amino acids is (almost) universal. Its universality is a very strong indication that all life on earth has evolved from a single source: in the words of Charles Darwin's grandfather, Erasmus, we are all descended from 'a single filament of life'.

It is now easy to see how misprints – mutations – among the genetic letters can have enormous consequences. For example, one of the most common single-gene defects in the world, the one responsible for sickle-cell disease which is prevalent in thousands of children of African origin, results from a change in just one letter in the sequence of base pairs specifying haemoglobin. Haemoglobin is made up from two kinds of amino acid chains, known as the alpha (α) and beta (β) chains, each of which is specified by its own gene. Two α chains and two β chains are present in each haemoglobin molecule. In the sixth codon of the gene specifying the β chain, instead of the sequence GTG, the sickle-cell variant specifies GAG. That is sufficient to alter the sense of the entire message – the RNA has no way of 'knowing' that there is a mistake in the message it carries out of the cell nucleus to the protein manufacturing machinery – and this in turn alters the structure of the haemoglobin molecule which the cell's machinery produces. The codon GTG specifies that the cell's protein manufacturers should put glutamic acid as the amino acid in the sixth position of the β chain, but those who carry the mutated codon GAG will have the amino acid valine in that position instead. This confers different properties on the haemoglobin, which in turn alters the shape of the patient's red blood cells so that they are prematurely destroyed in the circulation, giving rise to anaemia. The sickle-shaped cells are less flexible than normal with the result that they tend to clog up small blood vessels, cutting off the blood supply and causing serious damage to the tissues.

The solution to the problem of how cells make protein gives rise immediately to another puzzle. Every cell in the human body contains a complete copy of the entire human genome, but most cells are highly specialized: they are not all making all the proteins specified in the genome. For example, immature red blood cells (reticulocytes) make haemoglobin at a prodigious rate whereas some pancreatic cells make a very different protein, insulin, and do not 'express' the gene for haemoglobin at all. It follows that most of the human genome is actually switched off most of the time. This process of gene regulation is exquisitely subtle: a developing foetus, for example, will 'switch on' genes to make a special kind of haemoglobin, because of its high demand for efficient oxygen transport, and later in development will switch those genes off while activating the ones which code for adult haemoglobin. Relatively little is known about how genes are regulated. In the context of the Human Genome Project it is easy to forget the importance of this subject, but it is clear that knowing the genetic specification for all the proteins will be less than half the story. At least as important will be understanding how the action of the genes is orchestrated: in terms of the human body, the sum of the parts is less than the whole.

Throughout this chapter, there has been an unspoken assumption that genetic damage is solely an inherited condition. But this is not necessarily true. Many of the cells in an adult's body are dividing to replace those which have become diseased, damaged or which are simply worn out, and at each division the DNA is copied. Mistakes can arise in this process, despite the proofreading mechanisms which exist to prevent errors. Or a cell's DNA may sustain damage from radiation or chemicals in the environment. The altered genetic instructions are then passed on to all that cell's descendants. Someone who habitually smokes cigarettes exposes the cells lining their lungs to a cocktail of potent chemicals which can damage DNA; similarly, the skin of someone who spends a lot of time sun-bathing receives high doses of ultraviolet light, which also damages DNA. One result of this 'genetic' damage to body cells may be cancer. The rate at which cells grow and divide is determined by a very fine balance between genes promoting growth and those inhibiting it, and it may be that after some damaging event the cell slips out of normal control and starts to proliferate, ending up as a tumour.

The genetic disruption that gives rise to cancer may be at the level of the genes, but the damage is usually sustained in adulthood. The somatic or body cells of an adult are quite separate from the germ cells – the ones which give rise to sperm or eggs. Damage to DNA within somatic cells such as those in the lungs is confined to the organ within which the cell is situated and is not passed on to the germ cells and thus to subsequent generations. Only if, by some mischance, an embryo were to sustain damage very early in its development before the sex cells differentiated from the rest, could a defect be transmitted. In a very few human families, there are indeed cases where a strong predisposition to a particular cancer is passed down from parents to their children. These cases of inherited genetic damage can sometimes serve as models for much more common 'sporadic' cancers of the same type, because the inherited damage to DNA appears to be the same as that occurring in those cases of cancer resulting from environmental damage. By looking at inherited predispositions to cancer, researchers have already found important clues to how more common cancers arise. The realization that genetic damage underlies many cancers means that the most important applications of human genetics research may not lie with inherited disease at all but with cancer. Inherited diseases are comparatively rare: cancer is not. Whereas CF afflicts 1 out of 2,500 or so in the population, one out of four people in Britain will die of cancer.

One further important difference between the genetic damage

leading to cancer and that leading to single-gene defects such as CF, is that the development of a cancer requires the action of a cascade of different genes rather than just damage to one gene. Quite apart from the rare known 'inherited cancers', some people may inherit a more subtle predisposition to cancer: they may inherit from their parents damaged copies of some of the genes in the cancer-causing sequence, so are more likely to develop cancer as a result of environmental exposure than those who have inherited intact versions of the corresponding genes. From statistical studies of large populations, there is no doubt that smoking or chewing tobacco causes cancer, but there are many tales of people whose granny smoked forty Woodbines a day and lived to be ninety-five. It is currently impossible to predict of any one smoker whether they will develop lung cancer or not. One application of new knowledge about the human genome may be to pinpoint those individuals in the population who have inherited a predisposition for cancer. The idea of screening the population to find people at risk is not to pronounce a genetically preordained death sentence upon them. Multi-gene events, such as cancer, are not determined solely by one's inheritance but come about as the result of genes interacting with the environment. The idea would be to give at-risk individuals information to enable them to avoid dangerous environments – socializing in a smoky pub or working at the Sellafield nuclear reprocessing plant – and so prevent further genetic damage. Attractive though this prospect may be, there is a downside to such early diagnosis of risk which will be discussed in later chapters.

Modern genetics has already been applied to treat existing cases of cancer. In October 1991, Rosenberg and French Anderson, a year after carrying out the first successful gene therapy on ADA deficiency, began the first-ever attempt to treat cancer patients with genetically altered cells grown from their own tumours. They took tumour cells from a forty-six-year old man with a melanoma (a type of skin cancer) that had spread throughout the body and altered the cells to produce large quantities of a toxin called tumour necrosis factor. They hope this will make the tumour more susceptible to attack by the patient's own immune system. This gene therapy trial is much less straightforward than the case of ADA deficiency – the aim is more ambitious than just replacing a previously deficient chemical. At the time of writing, the results appear ambiguous.

Indeed, Rosenberg has been the subject of some criticism within the US medical community for not publishing enough information promptly, so that a full assessment can be made of the procedure's efficacy.

3

The Anatomy of the Human Genome

OVER the past thirty years or so, the American geneticist Victor McKusick has been compiling a list of all those human genes about which something, however slight, is known. Professor McKusick has devoted his professional life to the study of human inherited diseases, founding a department of medical genetics at Johns Hopkins University in Baltimore in 1957. Now, in extremely active retirement, he is universally regarded as the premier geneticist of man. His compilation of human genes forms a thick volume, *Mendelian Inheritance in Man*, and is perhaps the only publication in existence of which one can come close to saying, 'All human life is there'. It has just over 5,000 entries, less than one tenth of all the genes that go to make up a human being. No one knows precisely how many human genes there are, but estimates put the total number at between 50,000 and 100,000. Until the discovery of DNA the genes themselves could only be studied indirectly by tracking the inheritance of some trait down through the generations. This classical transmission genetics made little progress with humans, for *Homo sapiens* lives too long and breeds too slowly to be an easy animal in which to study inheritance. That geneticists have some information about even 5,000 genes is an astonishing tribute to the power of modern research.

From another perspective, McKusick's catalogue is a sad tale of human suffering. For much of our knowledge about human genes comes from the study of cases in which the genes are defective, give rise to disease, and thus identify themselves. Professor McKusick himself has referred to this aspect of human genetics as its 'morbid anatomy'. The human genome is formidably complex and, when examining any complex system, scientists will try and modify single components successively to see what effect that has on the whole. But one cannot modify a human being's genetic make-up simply to see what the result will be, and so the compilation of information about human genes has had to be based substantially on observing the effects of genes which have gone wrong in the natural course of events.

The discovery of the structure of DNA set a new agenda for genetics. Genes at last had a molecular and chemical embodiment which, at least in principle, could be dissected directly. Instead of tracking the transmission of characteristics which might skip generations before being expressed, geneticists could hope to analyse the chemistry of DNA itself. Latent in Watson and Crick's discovery was the ultimate goal of modern molecular biology, its Holy Grail: to read off the sequence of those chemical letters of the genetic alphabet – the Cs and As and Gs and Ts. And it is natural for us to want to decipher this sequence for what we, in our anthropocentric way, consider to be the most interesting of all living things – a human being. The agenda was set four decades ago; the Human Genome Project aims to accomplish the task by the early years of the twenty-first century.

The DNA in the cells of the human body is divided into 46 separate molecules, each packaged into a chromosome. A chromosome probably carries thousands of genes. Chromosomes are large enough to be viewed through a microscope once they have been stained to make them visible. The 46 chromosomes in human cells are made up of 23 pairs – one member of each pair is inherited from an individual's father and the other from their mother.

The business of modern genetics has become the chemical analysis of these molecules of chromosomal DNA. It is as if those engaged upon the Human Genome Project have wandered into a vast library, with the genes corresponding to the books, intent on reading every word of every book. There are certain essential preliminaries: if you are going to study a book in a library, it helps to have a catalogue. The basic structure of the library of the human genome is known, at least, in outline: it contains somewhere between 50,000 and 100,000 books, and almost every one is present in two editions. The library is divided up into 46 bays: 23 of them carrying one set of editions and the other 23 carrying the other edition of the same set of books. Thus, a geneticist who wishes to read the base sequence of a particular gene will also want to know where among the 23 pairs of chromosomes the gene is located. The process of building up an index or catalogue of the genes and relating that to their actual position on the chromosomes is called 'gene mapping'.

Mapping genes in human DNA began long before the Human Genome Project but progress was slow. As recently as 1956, no one even knew the correct number of chromosomes in cells of the human body. For decades researchers had accepted that humans had 48 chromosomes, and it was only in 1956 that Jo Hin Tjio and Albert Levan announced that the true number was 46. Improved methods for making human chromosomes visible under the microscope, which they and other researchers developed in the mid-1950s, yielded immediate benefits in understanding human genetic disorders. In 1959, the French cytogeneticists Jérôme Lejeune and Marthe Gauthier first showed that people with Down's syndrome had 47 rather than the normal 46 chromosomes.

So difficult was the business of separating and identifying human and other mammalian chromosomes that such an obvious question as how gender is determined was only worked out in the late 1950s. Cytologists (those who study cells) had known for a long time that males had a single X chromosome and a much smaller Y chromosome and that females had two X chromosomes. Studies on the fruit fly *Drosophila* suggested erroneously that it was the number of chromosomes which determined gender, leading geneticists to conclude that men were male because they had only one X chromosome and women were female because they had two. This idea that gender depended on the 'dosage' of a particular chromosome is correct for fruit flies but quite wrong for humans.

In 1959, the British cytologist Dr Charles Ford, working at the Medical Research Council's Radiobiology Unit near Harwell in Oxford, studied the chromosomes of women suffering from a condition known as Turner's syndrome. He showed this was associated with having only one X chromosome – such individuals lacked a second X, and they did not have a Y chromosome either. According to the prevailing theory they should have been male, because they had only one, not two, X chromosomes; in fact, all those with Turner's syndrome are female. At the same time, another British cytogeneticist, Dr Pat Jacobs, was studying men who had two X chromosomes and one Y. Since, by the dosage hypothesis, the individuals she was studying should have been female, the obvious fact that they were men showed that the presence of the Y chromosome determined male gender. This XXY genetic constitution is called Klinefelter syndrome and is surprisingly common: about one in 500 boys is affected in this way (the incidence for Down's syndrome, which is not sex-linked, is about one in 650 children). These pioneering studies on chromosome abnormalities demonstrated that the Y chromosome must carry a gene (or several genes) whose presence was necessary for the individual to be male.

But it was more than thirty years before geneticists identified the gene on the Y chromosome which acts as a 'switch' to tell the developing embryo that it is to develop as a male, and without which the embryo will naturally go down the path of becoming female.

They studied the very rare cases of XX men and XY women. The sophisticated DNA analysis techniques available in the late 1980s and early 1990s showed that in the XY women, their Y chromosome is broken or has had parts deleted and these missing stretches contain the crucial male-determining gene. The XX men appear to have inherited from their fathers an X chromosome which has gained a bit of extra genetic information – the male sex-determining gene – in a sort of swap with a paternal Y chromosome during the formation of the father's sperm. This chance event leading to the crossover of genetic material between an X and a Y chromosome during sperm formation is very rare: fewer than one in 20,000 men has an XX genetic constitution. In May 1991 a joint team from the Imperial Cancer Research Fund laboratories led by Dr Peter Goodfellow and from the Medical Research Council led by Dr Robin Lovell-Badge published definitive proof that they had indeed located the sex-determining gene. They had succeeded in switching the sex of three mice which had been conceived as females into males by injecting the newly fertilized eggs with a fragment of DNA containing the mouse version of the sex-determining gene.

Because men normally have just one copy of the X chromosome, if any genetic defects occur in men much more commonly than in women, the chances are that the defective gene is present on the X chromosome. If a woman inherits one defective copy of an X-linked gene, then she is unlikely to show symptoms of disease because she will most likely have an intact copy on her other X chromosome. There are very few active genes on the Y chromosome, so it figures very little in studies of transmitted characteristics other than maleness. A woman might inherit two X chromosomes each of which has a defective copy of the same gene, but the likelihood of such an event is much, much smaller than the chance that a man will inherit

43

a defective gene on his single copy of the X chromosome.

Perhaps the most celebrated example of an X-linked trait is haemophilia, which allegedly entered the royal families of nineteenth-century Europe through Queen Victoria and her many daughters: the defect can be transmitted through the female line, but women carriers are symptomless. It is their sons who manifest the disease. Their blood is unable to clot properly and they are liable to bleed copiously from the slightest cut and, more seriously, are liable to severe internal bleeding as a result of the sort of knock which would merely bruise another person.

The gene which, when defective, gives rise to haemophilia was not, however, the very first to be assigned to the X chromosome. Johann Friedrich Horner, a professor of ophthalmology in Zurich, started the process of mapping another gene. In 1876, he drew up a pedigree showing that colour blindness runs in families. In 1911, the American geneticist E. B. Wilson, working at Columbia University, assigned the gene for colour blindness to the X chromosome. Wilson had extended Horner's observations and noted that red–green colour blindness is much more common among men than women. He realized that some of the genes for colour vision must lie on the X chromosome because the effect of the defective variant was showing up in men. In the early years of the twentieth century, geneticists had thus assigned the genes for colour vision and for some blood proteins to the X chromosome. But no headway was made with mapping genes on the other chromosomes for nearly fifty years. By 1968, around 70 genes had been assigned to the X chromosome and this was the extent of the gene mapping which had been achieved by that time.

Because the Y chromosome, responsible for maleness, is the smallest chromosome of all, it can be easily distinguished from the rest through a microscope. Reliable ways of identifying and telling the other chromosomes apart from each other had to wait for technological developments which did not come until the late 1960s. Until then, the only dye available to stain chromosomes selectively made them visible as small red rods under the microscope with no discernible internal structure. For example, the X chromosome and seven of the non-sex chromosomes (autosomes numbers 6 to 12, as they are now identified) look virtually identical through the microscope. A start was made in 1959, when the so-called Denver

THE ANATOMY OF THE HUMAN GENOME

classification of chromosomes was introduced, which tried to arrange them according to their length and general appearance, assigning them numbers roughly in order of decreasing size; chromosome 1 is the longest and chromosome 22 was judged to be the shortest of the autosomes. In fact, the classifiers made a mistake, because subsequent work has shown that chromosome 21 is actually smaller than number 22. By the end of the 1950s, the gene mappers were in little better shape than medieval geographers for whom much of the earth's surface was *terra incognita*.

The limitations of the research laboratory were reflected in the clinical wards of the hospitals. Genetics was far from the forefront of medical practice during the immediate post-war period. In the developed countries, antibiotics were controlling or even eliminating the major infectious diseases which had killed and maimed millions throughout human history. Vaccination and antibiotics, together with the better diet and easier life-style which accompanied rising standards of living, meant that the broad mass of the populace in developed countries could expect to live longer and healthier lives than ever before in human history. While successive 'magic bullets' were doing their work and attracting money, prestige and public attention, human genetics was something of a backwater. However, as the incidence of the major infectious diseases has declined, so disorders with a genetic basis have grown in relative importance. In northern European populations, around 1–2 per cent of all births are associated with some form of genetic disability or with developmental disorders which have a strong genetic component. A study of the children admitted to a paediatric hospital in Montreal between 1969 and 1970 showed that 'genetic' admissions accounted for about 11 per cent of cases. Genetic disease is now a major cause of death in those under the age of fifteen.

In the developing countries, infant mortality due to infection and malnutrition has hitherto masked the burden of genetic disease. Yet it is in the developing countries that genetic disorders exact their heaviest toll, for the most common single-gene defects in the world – sickle-cell anaemia and thalassaemia – are those which arose as a side-effect of humans evolving a defence against malaria. In those areas where malaria has been eradicated, genetic disease has then been revealed.

Thalassaemia in Cyprus is one classic example. The thalassaemias

45

are a group of crippling disorders of the red blood cells, possibly the most common single-gene defect in the world, and their incidence is strongly correlated with the distribution of malaria. The condition was almost unknown in Cyprus until the early 1950s when malaria, previously endemic there, was eradicated from the island and death rates in early childhood declined. It then became apparent that thalassaemia had been present in the population but had never become manifest because those affected died young as part of the high infant mortality rate. The disease became a major public health problem on the island and it became clear that the cost of modern high-technology methods of caring for all the affected children – procedures involving repeated blood transfusions – would eventually eat up half the island's entire health care budget. Examples such as this have brought genetic disease to the forefront of modern medicine – for hard financial reasons as much as out of compassionate consideration for alleviating human suffering.

In 1970 came a real breakthrough in the physical mapping of human genes. Torbjörn Caspersson, Lore Zech and colleagues at the Karolinska Institute in Sweden found a way of 'staining' human chromosomes with fluorescent dyes to make characteristic banding patterns visible. When treated with the acridine dye quinacrine mustard, and illuminated with ultraviolet light, each chromosome became visible as a series of dark and bright bands. The banding technique most widely used today employs a fluorescent dye called Giemsa. The banding patterns allowed cytogeneticists to see the chromosomes clearly and tell them apart for the first time.

The technique of staining and examining chromosomes under the microscope is known as 'karyotyping'. Once each human chromosome could be identified by its banding pattern, researchers adopted a uniform system of naming the chromosomes and bands. Under the microscope chromosomes look like small rods pinched more or less near the middle, and this pinch, known as the 'centromere', divides each chromosome into a 'short' and a 'long' arm. In 1966 the short arm of each chromosome was designated arbitrarily as 'p' and the long arm as 'q' and in 1971, an international conference in Paris officially adopted a numbering system developed by Caspersson which divided each arm into up to three regions demarcated by very prominent bands. Regions are numbered out from the centromere to the end of the arm and bands are numbered consecutively within each region. Thus, for example, one of the genes for the protein collagen lies at 7q22: on the long arm of chromosome 7 in the second band of the second region. The cystic fibrosis gene is not very far away, at 7q31–q32, straddling the first and second bands of region 3. The gene which when defective gives rise to Huntington's chorea lies on the terminal band of the short arm of the fourth chromosome: 4p16. The genes for the ABO blood group lie near the tip of chromosome 9 at 9q34: on the long arm, in the fourth band of region three.

Karyotyping is a powerful way of screening prenatally for gross chromosomal abnormalities. The development of amniocentesis and chorionic villus sampling has made it possible to collect samples of cells which have the same genetic constitution as the developing foetus. In amniocentesis the doctor injects a needle through the wall of a pregnant woman's womb to draw out a small sample of the fluid surrounding the foetus. Although this fluid is not going to develop into the child, the cells in the amniotic sac are of the same genetic composition as the foetus not the mother. The cells can be grown in the laboratory, the chromosomes stained, and then counted and examined under the microscope. The problem with this technique is that amniocentesis is not possible until sixteen to eighteen weeks into the pregnancy and it takes a further week or so for the cells to be cultured so that the chromosomes can be analysed. If the mother of a foetus affected by a chromosomal abnormality wishes to terminate her pregnancy, it has to be done late, at around nineteen to twenty weeks, and this can be a deeply distressing procedure. More recently, a different technique, which takes a few foetal cells from the chorionic villus – a tissue which later develops into the placenta – has become available and this can be carried out much earlier, around the ninth to twelfth week. Chorionic villus sampling may carry a marginally higher risk of triggering a miscarriage than does amniocentesis, but for many women the benefit of early diagnosis may outweigh any disadvantages.

The most common chromosomal abnormality is the one leading to Down's syndrome, which is associated with the possession of three copies of chromosome 21 instead of two. Down's syndrome affects about 1 in 600 to 650 babies born each year. The likelihood that a woman might give birth to a Down's syndrome baby increases with her age from 1 in 2,000 for women in their early twenties to about 1 in

350 for women aged thirty-five. By the age of forty, the chances are 1 in 100 that the pregnancy will be affected by this abnormality. Once the karyotype has been made, following amniocentesis or chorionic villus sampling, the presence of an additional chromosome 21, the hallmark of Down's syndrome, is clear to a trained cytologist. Down's syndrome arises spontaneously and is not inherited from an affected parent (men with Down's syndrome tend to be sterile, although some women have had children).

The chromosome banding patterns provided, for the first time, a detailed physical cartography of the human genome. Researchers in the human genetic library could now not only tell one bay from another but, figuratively speaking, could recognize the individual shelves. When the staining techniques were first developed about 400 bands were obtained. Technical refinements mean that up to 3,000 different bands can now be identified in human chromosomes, but they do not represent individual genes – the length of chromosome occupied by any band is too long to be a single gene.

So what about the books occupying the shelves – the genes themselves? Research had already shown where a few books are located. Visible chromosomal abnormalities in patients with a genetic disease can sometimes suggest where the affected gene might be located. The study of 'extra' chromosomes had demonstrated that the book containing the recipe for maleness is contained in the bay marked 'Y chromosome'. Studies of the frequency of inheritance of single traits had shown that books on colour vision and some blood proteins are in the X chromosome bay. To get an idea of how close two genes might be to each other and of their order on the chromosome, geneticists turned to studying the way in which several traits are inherited together.

Before genes are transmitted from parent to offspring, they are shuffled during the process of preparing the egg in the mother and the sperm in the father. Sex cells differ from the cells in the rest of the body in that they only have one set of chromosomes: only when egg and sperm fuse does the fertilized egg acquire its full complement of 23 pairs of chromosomes. Sex cells are formed by a complicated process of division in which genes are shuffled and the chromosome pairs are separated. Before the chromosomes separate, the chromosomes of each pair line up alongside each other, each makes several crossovers with the other and they exchange pieces of chromosome so a child receives from its mother a chromosome number 1 which is genetically distinct from either of her chromosomes 1 but represents a selection from both. Of the genes present on the child's chromosome, some will be the variants inherited from his maternal grandfather and the rest will be the variants inherited from his maternal grandmother. The same process will have occurred as his father's sperm is formed. This process of 'recombination' which shuffles the genes into new combinations is one of the most important events in the process of inheritance. On average about 52 recombination events take place, between one and six on each chromosome depending on length, although it seems, for reasons not yet understood, that more recombination takes place when women's chromosomes are shuffled than when men's are.

Genes which are close to each other on a chromosome tend to cross over together – the closer they are to each other, the tighter this 'linkage'. Just as two cards that are close together in a pack stand a reasonable chance of staying together when the pack is shuffled, so the variants of two genes that are close to each other on a chromosome stand a reasonable chance of being inherited together. By studying patterns of linked inheritance of several characteristics, it is possible to assess their relative distance apart. Two genes on the X chromosome provide a good example of how the relative positions of genes can be determined in this way. One is the colour vision gene which when defective causes red–green colour blindness and the other is one that determines production of the enzyme glucose-6-phosphate dehydrogenase (G6PD). When this gene is defective only small amounts of G6PD are made – the condition known as G6PD deficiency. This first showed up when American soldiers started to be invalided out of the Korean War after unexplained haemolysis – destruction of their red blood cells. The geneticist Barton Childs, working at Johns Hopkins in Baltimore, traced the G6PD gene to the X chromosome in 1958 by the traditional route of noting that G6PD deficiency occurs in males but not females of affected families. G6PD deficiency is most common among male American blacks – where the incidence can be as high as 10 per cent – and Mediterranean Jews. If people with G6PD deficiency are exposed to certain chemicals or drugs, especially modern antimalarial drugs, their red blood cells tend to break down and they become anaemic. Some families will carry an

X chromosome with mutant copies of both genes and, by studying the frequency with which the two traits are inherited together and how often they become separated, it is possible to gauge how close the genes are to each other on the X chromosome.

So in the early 1960s, McKusick and two of his colleagues, Ian Porter and J. Schulze, set about testing black boys in Baltimore schools for colour blindness. Out of 3,648 tested, 134 were red–green colour blind and from them the researchers identified ten families in which some sons were both colour blind and had low levels of G6PD. Within these families there were other boys who were either colour blind or had little G6PD but not both, so it was clear that genetic recombination sometimes separated the two genes. The boys must inherit the genes responsible for the traits from their mothers. But it was impossible for the researchers to track the genes in the mothers who, being female, had two X chromosomes and so did not exhibit the traits. So the geneticists had to skip a generation and go back to the boys' maternal grandfathers. Among the families in which the maternal grandfather carried both defective genes, 19 out of 20 affected grandsons had inherited both colour blindness and low G6PD levels. The remaining one had inherited either one or the other. The fact that the two genes separated in only 5 per cent of the boys is a tight linkage, suggesting that they lie quite close to each other on the X chromosome, and this distance is given an arbitrary length – which represents the frequency of recombination between the two positions. If one then finds that the linkage between colour vision and another gene – A – on the X chromosome is weaker, then this suggests that A must lie further away. The order of the three genes could then be determined by finding the linkage between A and G6PD deficiency.

Genetic mapping of this type is an old technique which was first perfected in fruit flies (*Drosophila*) in the early twentieth century by American geneticists under the leadership of Thomas Hunt Morgan. By 1915, Morgan and his colleagues had located the relative positions of more than 85 genes in the fruit fly, using these methods. They found that there were four groups of genes which appeared to be inherited together – four 'linkage groups'. Since examination of fruit fly cells revealed that they had four chromosomes, this was powerful evidence that genes were located on chromosomes and the definitive book *The Mechanism of Mendelian Heredity*, published in 1915, first demonstrated the validity of the chromosomal basis of heredity. In honour of Morgan's achievement, the unit by which genetic distances are measured is called the morgan. In the example given above, the two genes which separate from each other only 5 per cent of the time are said to be 5 centimorgans apart. In general terms, along the human chromosomes, a genetic distance of 1 centimorgan corresponds to a physical distance of about 1 million base pairs. This is only a very rough and ready approximation as the relationship between genetic and physical distances can vary by a factor of five or even ten.

Fruit flies breed quickly, producing a new generation every fourteen days, so they are ideal for genetic studies. Humans breed more slowly, and unlike fruit flies it is morally impermissible to mate two people with interesting defective genes to see the effect on their children. Although humans can be difficult objects for genetic study because they live long and have few offspring, there is one advantage to studying inheritance among humans: no other animal keeps extensive records of births, marriages and deaths. Records live longer than people, so it is sometimes possible to reconstruct the transmission of a characteristic through previous generations to the present day just by searching the records. None the less, the study of human genetics proceeded very slowly. Until the late 1960s, virtually all progress had been with the sex-linked genes. Researchers were confined to only two bays of the human genomic library, the rest was shrouded in darkness. Was it possible to find any of the books in the other bays of the library?

The first assignment of a human gene to a chromosome other than X or Y – to an autosome – came in 1968, a few years before the detailed banding maps became available. Roger Donahue, then a research student starting work in McKusick's laboratory, did what all young researchers in this field did: he looked at his own chromosomes through the microscope. He found that one of his chromosomes 1 looked different from the norm. Chromosome 1 is the longest of all the chromosomes, but one of Donahue's looked as if it was becoming unravelled and was even longer than normal. This abnormal chromosome seemed to be unconnected with any abnormal trait or disorder, but Donahue decided to find out how he had inherited it and to see if there was any biochemical or physical trait that had been inherited along with it. If he could find something, the abnormal shape of the chromosome would stand as a physical marker for a gene carried on chromosome 1.

Together with Wilma Bias and the British population geneticist James Renwick, he searched for variations in blood groups that might have been inherited along with the unusual shape of the chromosome. In addition to the well-known grouping A, B, AB and O, which hospitals need to determine to avoid rejection of a blood transfusion, human blood can be categorized into numerous other groups. The groups are all characterized by the different proteins carried on the surface of the red blood cells, which can vary from individual to individual. It turned out that every member of Donahue's family who carried the uncoiled chromosome 1 also carried a gene for a particular type of blood group known as Duffy-a. The assignment of the gene for Duffy-a to chromosome 1 was confirmed by studies of other families which also carried this chromosome abnormality. It was the first assignment of a gene to an autosome in humans. Within months, a group working with Renwick had assigned another gene to chromosome 16, to be followed rapidly by a different American group who identified a second gene with chromosome 16. In both cases they made the assignment by finding a correlation between a particular trait and a visible irregularity in the structure of the chromosome.

While the cytogeneticists were studying the physical topography of chromosomes under the microscope, a powerful method of isolating human chromosomes had become available. In 1967, Mary Weiss and Howard Green at New York University refined a technique for fusing human and mouse cells grown in culture. Some of the human–mouse hybrid cells would grow and multiply in culture, although they tended to throw off some of the human chromosomes as they went through the process of cell division and replication.

None the less, Weiss and Green were able to establish stable lines of hybrid cells containing the full mouse genome and a few human chromosomes as well. Each hybrid cell line contained a different subset of between eight and twelve human chromosomes in addition to the mouse chromosomes. (It is easy to tell mouse and human chromosomes apart under the microscope.)

Using a large set – a panel – of these 'somatic cell hybrids' containing different combinations of human chromosomes, it was then possible to correlate the presence or absence of a particular chromosome with the presence or absence of a particular biochemical trait – and thus of a gene. Biochemists can find the human protein that is produced by a given cell line and then look through all the different cell lines to find the human chromosome common to those cells producing the protein. In this way they can correlate the presence of the protein with the presence of a particular chromosome. In their very first paper on the topic in 1967, Weiss and Green noted that the enzyme thymidine kinase was related to a specific human chromosome, but had no means then of identifying it. Chromosome banding solved that problem, revealing exactly which human chromosomes remained stable in a given line of mouse–human hybrid cells. And so, four years later, Barbara Migeon and C. S. Miller at Johns Hopkins were able to identify chromosome 17 as the one carrying the thymidine kinase gene. Hybrid cell lines are now available that contain single copies of certain human chromosomes – 7, 16, 17, 19, X and Y for example – making the task of assigning a gene much easier.

So, by the start of the 1970s, cytologists and geneticists had begun to develop the maps necessary for finding their way round the human genome library. The maps were of several kinds and differed in fineness of scale. The most obvious step is to try to find out which chromosome carries a given gene – in which bay of the library does the book reside? If two genes are known to be on the same chromosome, one route to fixing a gene's position, the genetic linkage map, is to locate it by its proximity to others. In a library ordered by author's surname, *Anna Karenina* by Tolstoy will tend to lie close to and on the right-hand side of *The Hobbit* by Tolkien, and they will be closer to each other than either is to *Vanity Fair* by Thackeray. A genetic map is rather like the author index catalogue of a library. All the cards in the card index may be arranged in order but this does not tell prospective readers where physically they have to go to find the volumes themselves. Similarly, a genetic linkage map does not contain precise information about where among the sequences of base pairs on a given chromosome the genes reside. A physical map of a library would show on what shelf and how far along the shelf a particular book was located.

The bands which Caspersson's dyes revealed along the chromosomes provided the first means of measuring such distances in the human genomic library. In June 1973, Frank Ruddle of Yale University organized the first workshop on human gene mapping in New Haven. The Howard Hughes Medical Institute has published a short history of gene mapping, compiled by science writer Maya Pines, who records that by the time of the New Haven meeting 'only 150 genes were mapped to specific chromosomes, and Thomas Shows of the Roswell Park Memorial Institute in Buffalo New York, chairman of the nomenclature committee, could remember the precise name and symbol of each.' The entire human gene map consisted of a schematic drawing of each chromosome, showing the bands and the approximate location of each of the 150 genes. It took up no more than one A4 page.

The revolution arrived in the 1970s, with the invention of recombinant DNA techniques – the discovery of chemical scissors which could snip the double helix strand with exquisite precision in specific places and, further, of biochemical 'Scotch tape' which could bind two different cut strands together. Here at last were molecular scale 'scalpels' with which to dissect DNA, and which were so successfully applied to the treatment of human disease at NIH in September 1990. The era of recombinant DNA has revolutionized all genetics, including the study of human inheritance. Within the past decade gene-splicing techniques have been applied to making detailed maps of the human genome. By 1987, at least 1,215 human genes had been mapped to particular chromosomes. And the pace is accelerating. By October 1991, 2,144 genes had been mapped. The techniques which made this revolution possible are described in the next chapter.

4

From Microbes to Men

IT is not every day of the week that a university professor tells you to spit in front of him. But one morning, at St Mary's Hospital Medical School in London, Bob Williamson, the professor of molecular genetics, asked me to do just that. He had handed me a leaflet about the inheritance of cystic fibrosis (CF) and spent a couple of minutes running through a standard genetic counselling speech. I then took a mouthful of sterile saline solution from the sealed specimen bottle he gave me, rinsed my mouth out and spat back into the bottle. I had just contributed enough of my own DNA for Professor Williamson's research assistant, Ed Mayall, to dissect my genes.

Professor Williamson and his team have taken samples that way from nearly 3,000 people. The idea is to identify couples who each carry a mutated gene and who are therefore at risk of having a child who suffers from CF. Such a programme of genetic screening is only possible as a result of the exquisite refinement of today's DNA analytical technology. From the entire library of my 50,000 or so genes, Ed Mayall can pluck out and read the book for CF. As I rinsed my mouth with the salt solution, a few cells from the inner lining of my cheeks got washed out and deposited in the specimen bottle. In less than four hours, Mayall had separated the cells from the rest of the solution, teased out the DNA from the cells, copied it many thousands of times, and then matched my DNA against reference DNA samples containing mutations that cause CF. Parts of the procedure are semi-automated so he can process several samples at once, managing anything from 250 to 500 samples a week if need be. From the point of view of those taking the test, the procedure is simpler than giving a blood-alcohol sample for drink-driving. In my own case, the result was a resounding verdict of ordinariness. Neither the maternal nor the paternal version of my gene is the CF variant: I am one of those 24 out of 25 members of the population who do not carry the CF mutation in my genes.

This test is just one result of the technological revolution over the past twenty years. The revolution started in an unlikely place – the human gut. There are to be found the powerful keys that will unlock the human genome's secrets. These keys are contained in organisms which, at first sight, have little relevance to human genetics – bacteria. The study of bacterial genetics, starting back in the 1940s, has spawned a multibillion dollar industry, and transformed scientists' understanding of the genes of every living organism – including human beings.

The most intensively studied microbe is a bacterium called *Escherichia coli*, or *E. coli* for short, which inhabits the human gut in huge numbers. It is a remarkably successful organism; there are more *E. coli* in the gut of each individual human being than there are human beings on the surface of the planet. Cells of *E. coli* are tiny rods about two millionths of a metre (2 μm) long and half that across. It is however a very convenient organism for a geneticist to study: it is single celled, multiplies quickly – under ideal conditions one cell will divide to make two every 20 minutes – and will thrive in a watery solution containing just glucose and a few mineral salts.

Bacterial DNA is organized into a single circular 'chromosome' which lies free in the cell, unlike the much more complex cells of fungi, plants and animals, where the chromosomes are locked up within a special spherical compartment called a nucleus, which is isolated by a membrane from the rest of the cell.

In the late 1930s and early 1940s, before the role of DNA as the genetic material had even been proved, a group of researchers in the United States, among them several émigrés from Europe including Salvatore Luria and the physicist turned biologist Max Delbrück, realized the advantages for the study of genetics in both *E. coli* and the viruses – *bacteriophages* – that multiply in it. Only by working on one of the simplest organisms, Delbrück felt, would scientists be able to understand at a fundamental level the workings of genes. These researchers became known as 'the phage group', and in the late 1940s the young Jim Watson became a member.

The choice of bacteria and phage proved a remarkable piece of foresight and brought scientific dividends as important as those from Morgan's choice of fruit flies for the study of classical genetics. When Hershey and Chase demonstrated conclusively in 1952 that DNA was the molecule of inheritance, they used bacteria and phage. For more than three decades, from 1940 to 1973, bacteria and phage proved to be the only convenient systems for studying the organization and function of genes in any detail, and the consequences of introducing new genetic material into a cell.

Consequently, most of the advances in genetics came as a result of studies of bacteria and their associated phage. It turned out that bacteria have sex – that there are, so to speak, males who can pipe part of their DNA across to a recipient female. Researchers were able to interrupt the process when only some of the DNA had been transmitted and compare the effect with bacteria that had received more DNA – in effect, they engaged in selective breeding of the bacteria. Bacteria grow and reproduce rapidly, and they have many fewer genes than more complex organisms such as animals or plants, so the experiments were comparatively easy and quick to perform and the results were more clear cut than was possible with plant or animal genetics. It was even possible to study the order of genes around the bacterium's circular chromosome. The relative ease with which a bacterium's genes could be mapped by such means forms a strong contrast to the difficulties, detailed in the previous chapter, which faced researchers trying to map the human genome. More than 900 genes have now been mapped in *E. coli*, but this is probably still fewer than half of all the bacterium's genome. Although researchers thus accumulated a great deal of knowledge about how bacterial genes were organized and how they worked, they did so mainly by indirect 'genetic' means – no one had yet found a way of looking directly at the sequence of base pairs of any piece of DNA.

At the end of the 1960s things were poised for change. For our purposes, the intimate connection between microbes and men lies in the discovery of a battery of enzymes, produced by bacteria and viruses, which can be used to manipulate DNA. Gradually, almost without realizing it, researchers began to assemble a powerful toolkit with which they could 'edit' the genetic text written on the double helix. They discovered molecular scissors with which they could cut out a fragment of text, and biological Scotch tape to paste the edited text together. Most important of all, the realization dawned that the microbial editing kit for DNA was capable of highlighting specific texts within DNA. It was as if one could find the text of Hamlet's famous soliloquy in the library by pulling out all chunks of text

57

containing the phrase 'to be', and then picking out the pieces belonging to the soliloquy. The discovery of these editing tools and of how they could be used has made the dream of mapping and sequencing the human genome possible.

The technological revolution in molecular biology started around 1970. Hamilton Smith from Johns Hopkins University at Baltimore isolated from the bacterium *Haemophilus influenzae* an enzyme which cut the strands of DNA. Enzymes that cut up DNA at random had been known for some time, but Smith's enzyme was different. These molecular scissors had the remarkable property of recognizing a specific short sequence of base pairs and then always snipping the strand in the same place within that stretch. This was the genetic equivalent of the 'text search' facility on a word processor – the biological machinery for the equivalent of examining a whole library and cutting the books apart whenever the phrase 'to be' appears. The specificity of these restriction enzymes, as they are known, is astonishing. One restriction enzyme (*Eco*RI) produced by *E. coli* recognizes only the sequence GAATTC and always cuts between the G and the first A. The complementary strand of the double helix will have the sequence CTTAAG and *Eco*RI also snips this strand between the G and the A, making a staggered cut across the DNA. Hundreds of such enzymes, each with a different cutting site, are now known.

The key realization in the early 1970s was that these restriction enzymes will snip the DNA of any other cell just as effectively anywhere their particular restriction site occurs. So they can be applied to the study of the DNA of plants and animals.

By the early 1970s, routine methods had also become available for sorting DNA fragments by length. Under the influence of an electric field, strands of DNA will move through a jelly made out of a material known as agarose. Smaller DNA strands will move faster than larger ones. This technique, known as gel electrophoresis, will therefore sort DNA fragments according to their length and distinguish between fragments of different lengths produced by the cutting action of a restriction enzyme. For the first time, geneticists had a way of picking out a specific stretch of DNA from a jumble of fragments.

Bacteria also offered something more than just a mechanism for isolating a stretch of DNA, they also provided a way of molecular 'Xeroxing'. Geneticists could now 'clone' a human gene, by 'cut and paste' methods. A strip of DNA isolated by the cutting action of a restriction enzyme can be spliced into a bacteriophage chromosome to make a 'recombinant DNA'. *E. coli* cells can then be infected with the genetically engineered phage. The virus reproduces itself within the bacteria, faithfully copying the strip of human DNA along with its own, until so many new virus particles are produced that the bacterial cells burst open. The viruses are separated from the cell debris, the phage DNA isolated, and the human DNA recovered from it by cutting with the same restriction enzymes used to cut it out of the human DNA in the first place. In this way, at long last, geneticists were able to obtain a pure sample of any piece of DNA in large amounts for further analysis and use.

The business of genetics was transformed by the availability of pure DNA in quantities large enough for analysis. No one studies a chemical reaction between two individual molecules because they are too small to be readily observable and the energies involved in the reaction are tiny; instead, chemists amplify the effect up to a measurable level by having millions of reacting molecules in their test tubes. The difficulty for molecular biologists who wish to study aggregates of many millions of DNA molecules has always been obtaining a 'pure' sample consisting of many copies of only that piece of DNA wanted for analysis. The chemical scissors of restriction enzymes, with their astonishing specificity, and the DNA photocopying machinery of phage and bacteria, began to address that need.

The first recombinant DNA was constructed in 1970–71 by Paul Berg at Stanford University in California together with David Jackson and Robert Symons. Using restriction enzymes to cut the DNA and bacterial ligases to rejoin the cut ends, they spliced a piece of bacterial DNA into the DNA from a small animal virus, to produce an entirely new 'chimaeric' DNA – a ring of DNA formed from genetic material from two different sources.

Constructing a chimaeric DNA was all very well, the next problem was to find a way of multiplying it up into larger amounts. In 1973, Stanley Cohen at Stanford University in California together with Herbert Boyer at San Francisco showed how this could be done. They constructed a recombinant DNA from pieces of two different bacterial plasmids – small circular DNAs found in many bacteria – and reintroduced their recombinant plasmid into *E. coli* cells. As the bacteria multiplied, the recombinant DNA was multiplied along with all the rest of the bacterial DNA. By the following year, they had made recombinant DNAs including transferring DNA from an animal – the African clawed toad *Xenopus* – into *E. coli*.

At this point, worries began to surface about the propriety and safety of what was going on in the recombinant DNA laboratories. Was it right to patch genetic material from complex animals such as the African toad into the DNA of a bacterium such as *E. coli*? Was this not, in some way, a form of miscegenation? But the safety arguments had still greater strength. *E. coli* is ubiquitous and an inhabitant of the human gut. Some of the proposed experiments involved stitching *E. coli* together with SV 40, simian virus 40, which causes cancer by insinuating its DNA into the genes of host cells. Were the genetic engineers about to unleash a devastating plague of mutant cancer-causing bacteria? These concerns were widely articulated within the scientific community itself. In July 1974, a committee headed by Paul Berg called for a moratorium on genetic manipulation until the possible risks had been fully investigated. In February 1975, a group of some 150 of the foremost molecular biologists met in the Pacific coast resort of Asilomar to debate the issues. They started the process of devising rules to ensure the safety of recombinant DNA research. The rules have since been developed and incorporated into legislation in most of the scientifically advanced nations.

Geneticists now realized that they could transfer DNA into an unrelated cell and see what effect the transferred gene had – what protein it produced, for example. In effect, the researchers in the 1970s had reinvented sex. But this time there was a difference. Throughout evolution, sex – the transfer of genetic material from one cell to another – has usually been possible only between individuals of the same species. But now, with the panoply of techniques invented in the 1970s, biologists could take, for example, the genes for human insulin, put them into *E. coli*, grow up the bacteria in culture and harvest pure human insulin which the *E. coli* had been instructed to make. These techniques of recombinant DNA, or genetic engineering, are now the basis of a great industry. In manufacturing plants across the developed world, tanks of bacteria or yeast are busily brewing up rare and valuable pharmaceuticals, many of them based on recipes taken direct from human DNA. Today the biotechnology industry represents the most visible application of the techniques of recombinant DNA, as bacteria, yeasts, and even some mammals, are put to work making protein according to a genetic recipe that is not their own.

As the Human Genome Project isolates and identifies more and more previously unknown human genes, the potential for harnessing microorganisms to make scarce human proteins will grow enormously. As abundance replaces scarcity, moral questions will arise about what are the right choices in using the compounds plentifully available as never before. It is therefore helpful to sketch out here how the technology works.

The first human gene product to be produced in bacteria was the hormone somatostatin. This acts as a brake on growth of the body, counteracting the effects of insulin and of human growth hormone. People who have a deficiency of somatostatin often grow abnormally tall and the ends of their bones are abnormally large. The human hormone was virtually unobtainable and the endocrinologists who worked out the chemical structure of the equivalent animal protein had to process nearly half a million sheep's brains to extract five thousandths of a gram of material. Human somatostatin is a small protein, only 14 amino acids long. In 1977, Keiichi Itakura, Herb Boyer, Francisco Bolivar and colleagues announced that they had persuaded *E. coli* to make it in large quantities.

The most striking thing about their achievement is that they had synthesized the gene themselves rather than using a somatostatin gene cut out of a living cell. From a knowledge of the protein's amino acid sequence and of the genetic code which translates the language of DNA into that of amino acids, it was simple (at least in principle) to manufacture the string of nucleic acids which would direct synthesis of the protein. In fact, the process was a very sophisticated one chemically. The researchers then stitched the artificial gene into a plasmid and got that incorporated into the bacteria. The research was funded by one of the first biotechnology companies, Genentech, which Boyer and Itakura had set up.

Two years later, the company's scientists announced that they had successfully produced human growth hormone – somatotropin – in bacteria. This was technically more difficult because the protein is 191 amino acids long. This time the researchers synthesized only part of the gene chemically and married up what they had made with part of the naturally occurring gene. Although more difficult to produce, human growth hormone is potentially more lucrative commercially than somatostatin. It is used to treat children who suffer from growth hormone deficiency and who would, without injections of the hormone, be stunted in their growth and end up as dwarfs. In addition, it is used to help broken bones to join, to mend burned skin and to heal bleeding ulcers. There is a controversial suggestion that it might slow up some of the processes associated with ageing.

Genentech started manufacturing its genetically engineered human growth hormone long before there was any notion of researchers embarking on the Human Genome Project, but the way in which society has dealt with the plentiful supply of a previously rare and expensive human protein is an important pointer to the potential consequences of the flood of new proteins which will be cloned and mass-produced as a result of human genome sequencing. The issues raised will be discussed in Chapter 7. The commercial market for growth hormone and somatostatin is limited because dwarfism and gigantism are fairly uncommon conditions, but in a sense these proteins were but models on which researchers tried out and perfected their techniques in preparation for cloning and manufacturing insulin, the biggest commercial prize. Human insulin manufactured by genetically engineered bacteria was duly authorized for the treatment of diabetes in humans in the early 1980s. It was widely welcomed as a better alternative to the pig's insulin, extracted from the pancreas of pigs at the slaughterhouse, that had been used hitherto. However, the use of recombinant human insulin has recently been the subject of some controversy, with reports of allergic reactions and of diabetics finding it difficult to gauge the dose required to maintain their blood sugar at its proper level.

At the same time that genetic engineering was getting under way, the geneticist's ability to isolate human genes was steadily increasing.

In August 1978, a team of researchers at the California Institute of Technology, led by Tom Maniatis, announced that they had created a human gene 'library' by cutting up human DNA with restriction enzymes, inserting the fragments into bacteriophage vectors and growing these in *E. coli*. Maniatis and his colleagues obtained their DNA from the liver of a miscarried human foetus. They started with thousands of copies of the entire human genome. They then treated the DNA with two restriction enzymes in such a way as to get DNA fragments about 20,000 base pairs long. Using now-standard recombinant DNA techniques, they then stitched each fragment into the DNA of their chosen vector phage and infected *E. coli* with these recombinant DNAs.

They had manipulated the original digestion conditions so that their eventual library of around a million fragments should cover the whole genome. The viruses dutifully reproduced themselves within the bacteria, copying the human DNA in the process. By standard culture techniques, phage derived from each individual recombinant phage were isolated and stored. The hitherto intractable human genome was now available as a million small fragments, each conveniently packaged so that it could be individually picked out and analysed. This and many other human DNA libraries that have been made since are the raw material with which the Human Genome Project works.

Calling the collections of cloned DNA 'libraries' is something of a misnomer, however, although it has now passed into common usage among molecular biologists. Restriction enzymes just cut human DNA up into fragments without offering any means of identifying even from which chromosome they might have come. They do not provide a catalogue or reference system. Before one can make full use of the library one has to find out what genes each fragment contains and in what order the fragments fit together to make up the sequence of DNA along each chromosome. To address the latter problem a 'physical map' of the genome is needed. This will be a very different map from the chromosome banding patterns discussed in the previous chapter. One approach, known as a contig map, capitalizes on the fact that the fragments produced by a partial digestion of the human genome with restriction enzymes will overlap with each other – contiguous overlapping fragments are known as 'contigs'.

Conceptually, contig mapping is like photocopying a library many times, dividing up each duplicate library virtually at random and then comparing all the different piles of books thus created from all the libraries to see which have some volumes in common. In practice, many technical difficulties remain in constructing even partial contig maps to cover the human genome.

Maniatis's original library contained about a million clones: it is clearly formidably difficult to sort through a million fragments and try to arrange them in sequence. The vectors – the carrier DNAs –

which were first used to make gene libraries would accept only comparatively short stretches of DNA: up to about 50,000 base pairs. In 1987 Maynard Olson and his colleagues working at Washington University in St Louis found a way round this. They developed a way of packaging several hundred kilobase pairs (thousand base pairs) of DNA into an artificial chromosome which would replicate itself in yeast cells. These 'yeast artificial chromosomes' (YACs) have become the workhorse of all DNA mapping, for they allow the division of large genomes into comparatively few fragments, which makes easier the task of sorting them. Their development was one of the technological advances which has made mapping and sequencing the human genome possible.

Crucial to mapping genomes is the ability to distinguish one DNA fragment from another. The double-stranded structure of DNA itself offers a way of identifying fragments. A single strand of DNA and its complementary partner will pair up with one another to re-form the double helix in a process known as 'hybridization'. A single strand of DNA can thus act as an exquisitely sensitive 'probe' to seek out a complementary strand within a jumble of DNA. The probe will ignore fragments which have a different sequence of base pairs. Methods of identifying DNA fragments using this principle permeate all recombinant DNA technology and physical gene mapping techniques.

Obtaining the probe that will pick out a particular gene is the most difficult part of the process. In a few cases, it is relatively easy. For example, developing red blood cells do little else but make haemoglobin. This means that the haemoglobin genes are being copied at a frantic rate into messenger RNA. Any RNA extracted from the cell will contain large amounts of messenger RNA corresponding to the α- and β-globin genes. This can be separated out, radioactively labelled, and used as a probe directly. But far more useful is the technique of copying the messenger RNA (mRNA) into DNA first, using the enzyme reverse transcriptase. This enzyme – discovered independently by Howard Temin at the University of Wisconsin and David Baltimore at MIT – takes a single-stranded RNA as a template upon which to construct a complementary strand of DNA. It comes from retroviruses, a type of RNA virus which needs the reverse transcriptase in order to make a DNA copy which is insinuated into the host cell's DNA.

64

When mRNAs extracted from a cell are mixed with reverse transcriptase and the required nucleotides – the building blocks of DNA
– a complementary DNA strand is built up along each RNA. If
radioactive nucleotides are used a radioactively labelled DNA is
obtained which can then be used as a probe. Converting mRNAs into
cDNA in this way also enables the genetic messages they contain to
be cloned, just like any other DNA. So the original mixture of
mRNAs extracted from a cell can be converted en masse into cDNA
and then the individual types separated out and multiplied up into
large amounts by cloning.

If at least something is known about the amino acid composition of
the protein for which the gene codes, then it is possible to make a
gene probe artificially, by applying the genetic code in reverse:
deduce the sequence of codons in the gene from the sequence of
amino acids in the protein. This approach too has its drawbacks since
in general more than one codon can give rise to one amino acid,
although there are ways of minimizing this problem. Often, however,
nothing is known about the sequence of the protein. One ingenious
way of getting round this difficulty is to use the exquisite sensitivity of
the immune system to raise antibodies against the protein, use the
antibodies to pick out the protein while it is still being made, and
therefore isolate the complex consisting of the part-formed protein,
its messenger RNA, and the protein-making machinery. The
mRNA can then be separated out and used as the basis for making a
cDNA probe. There are other even more complex and ingenious
methods, but it still remains the case that detecting a specific gene in
the 3 billion base pairs of human DNA is a formidable task.

The identification of DNA probes associated with specific genes
allows researchers to locate the region of a chromosome in which the
gene resides. Once a large quantity of the DNA is available, the
hybridization technique can identify the chromosomal location of a
gene directly. Probes containing the gene are mixed with denatured
chromosomes – the mouse–human somatic cell hybrids described in
the previous chapter offer one convenient way of providing individual
human chromosomes for this 'in situ hybridization'. Once again, if
the probes are radioactively labelled, researchers can tell from an
exposed X-ray film which human chromosome the probe has bound
to. More recently, fluorescent chemicals have been used to highlight
the position of the probe, which is quicker and more convenient than

radiography. The technique ties in with the chromosome banding discussed in the previous chapter, for it is possible to tell within which of the chromosome's bands the DNA probe has hybridized, thus filling in more details of the large-scale physical map of the human genome.

All the physical maps mentioned so far are fairly large scale and can only approximately locate a gene or the stretch of DNA associated with a probe. Ultimately, the highest resolution physical map of the human genome is the sequence of base pairs itself. The late 1970s saw a vital technical advance which makes this possible. In 1977, Fred Sanger in Britain and Walter Gilbert and Allan Maxam in the USA independently developed methods for rapidly reading the sequence of base pairs in any stretch of DNA. This had been the dream of geneticists for decades, ever since the structure of DNA had been uncovered and the genetic code cracked. But no one could start to address the problem of rapid and easy sequencing of long stretches of DNA until they could get a reproducible supply of small identifiable fragments that they could eventually piece together. Restriction enzymes, gel electrophoresis and DNA cloning provided them.

In 1977, Sanger and his colleagues in Cambridge announced that they had sequenced the entire DNA, all 5,577 base pairs, of a phage known as ϕX174 using their method. It was not, in fact, the first to be entirely sequenced. The previous year, a group of Belgian researchers had finished the RNA sequence of another phage – MS2 – but the Belgians had spent some fifteen years doing it, and had had to use incredibly laborious methods. According to an editorial published in 1980 in the journal *Nature*, the methods developed by Sanger and Gilbert meant that 'the sequencing of MS2 would now be no more than a single PhD project'. Sanger and Gilbert were each awarded a quarter share in that year's Nobel Prize for chemistry, with Paul Berg getting the other half. (Sanger had already been awarded the 1958 Nobel Prize for his work on insulin; in 1980 he became one of the very few scientists to have won the Nobel Prize twice.) By then, DNA sequencing was commonplace; by the mid-1980s it had become automated by Leroy Hood at California Institute of Technology. Nowadays machines are commercially available and are standard equipment in any modern laboratory engaged in DNA analysis. At present, major efforts are being devoted to mapping and sequencing the genome of *E. coli* (about 4.7 million base pairs), the fruit fly *Drosophila* (some 165 million base pairs long), and the nematode worm, *Caenorhabditis elegans* (whose genome is about 100 million base pairs). One reason for looking at these simple genomes is as models of how to carry out sequencing on the far more complex human genome.

In the mid-1980s, a radical new technique became available for making billions of copies of a single piece of DNA, without the bother and complexity of genetically engineering it into a vector and multiplying it in bacteria. Surprisingly, the new technique – known as the polymerase chain reaction (PCR) – made use of an enzyme, again taken from bacteria, which had been discovered nearly thirty years previously and whose potential had not been fully realized until the 1980s. In 1956, at Washington University in St Louis, Missouri, a group led by Arthur Kornberg isolated from *E. coli* an enzyme, now called DNA polymerase I, which will assemble a DNA double helix in the test tube, by attaching to a single strand of DNA the complementary bases which make up the companion strand. (The Kornberg enzyme is not the one, however, responsible for duplicating the DNA double helix when a cell is dividing: it is too slow, and other enzymes exist to do this.) All that is required is that the sequence of base pairs flanking the segment of interest is known, so that the scientist can direct the copying process to start in the right place. It was this technique that Ed Mayall used in the test to see whether I carried the mutation for CF. He first spun the saline solution which I had spat back into the sample bottle at 4,000 revolutions a minute for five minutes in a centrifuge. This collected all the solid debris – containing a few thousand cells from the inner lining of my cheek – into a discoloured spot at the bottom of the tube. He then rinsed this, redissolved it in saline solution, and boiled it for half an hour in sodium hydroxide solution. This rather drastic treatment has the effect of breaking open the cell and releasing the DNA. Half an hour at about 90°C denatures the DNA – separates the twin strands of the double helix molecule from each other. Using a diagnostic kit that incorporates PCR, he then made hundreds of millions of copies of the CF gene in my DNA.

The procedure takes about four hours and is simple: add the appropriate biochemicals and put the mixture in a special machine. The chemical reagents cost a couple of pounds sterling and the machine is about half the price of a commercial office photocopier. PCR takes a single strand of DNA as a template and rebuilds the double helix by stitching together a complementary strand. The double helix is then denatured by heating and the two new single strands each act as templates for two new helices. In each cycle the amount of DNA doubles. To overcome the problem that the heat needed to separate the DNA strands also activated the Kornberg polymerase and fresh enzyme had to be added at each cycle, scientists at Cetus, where PCR was first developed, hit upon a very neat solution. They now use DNA polymerase isolated from a bacterium that lives in hot springs, as this polymerase remains stable and active at high temperatures.

To start the reaction, Mayall took half my DNA sample and added some 'primers' – short stretches of DNA, made by Cellmark, the makers of the kit, and whose base sequence is identical to part of the CF gene. These primers stick to the appropriate place in my sample of single-stranded DNA and PCR then adds new bits of DNA to the end of the primer until a complementary copy of the gene has been completed. If I carried a mutation, the primers for the intact gene would not bind to my DNA properly and the reaction would fail. Into the second sample, Mayall added primers containing a known mutation: if I carried the mutation, the reaction would proceed but if I had the intact gene it would fail.

After the machine had gone through 35 cycles of heating and cooling we had hundreds of millions of copies of my CF genes in the tube. Even so the test tubes looked no different to the naked eye. At this point, the technique of gel electrophoresis was brought into play. An hour later we had a pattern of DNA fragments which would reveal the presence or absence of any mutations. Compared to the halting and crude methods of analysing human chromosomes which had been available for so much of the post-war period, PCR is a powerful technology indeed. The moral problems of harnessing this power and testing the population at large for genetic defects in this way will be discussed in Chapter 6.

One of the first discoveries when the new recombinant DNA based techniques were applied to studying DNA was that the human genome (like many others) is much more complicated than had been supposed. In 1977, while studying adenovirus – a virus which causes respiratory infection in humans – two teams of researchers, one led by

Phillip Sharp at MIT and the other by Richard Roberts at the Cold Spring Harbor Laboratory, discovered that its genes were interrupted by 'intervening sequences' which were not translated into protein. The news amazed the scientific community at first, but then introns, as these intervening sequences were soon called, began turning up everywhere: in other viruses, and in the DNA of chickens, rabbits and humans. The only class of living organisms in which split genes are virtually unknown is the bacteria. Bacteria on the whole pack their genes neatly, with every base pair playing its role as a letter forming part of the instructions for making a protein. The DNA of plants and of animals is much more prodigal. Within one gene there are usually one or more introns which have no protein-coding function and which the cell's biochemical machinery has to edit out before it can translate the gene into a protein (the blocks of DNA containing coding instructions are now known as exons). When a gene is expressed, the entire gene is copied into RNA, but the meaningless parts are spliced or edited out, to provide a functional mRNA that the protein manufacturing machinery can translate.

The total length of the introns within a gene can be much longer than the length of the coding regions. The gene which contains the instructions for adenosine deaminase (ADA) spans about 32,000 base pairs. But the actual recipe for ADA, the coding regions of the gene, are only about 1,500 base pairs long in total. There are eleven introns interrupting the coding sequence. And the ADA gene is medium-sized compared to some: the gene implicated in Duchenne muscular dystrophy spans more than 2 million base pairs of DNA, even though the coding region is probably only about 17,000 base pairs long. There are around 50 introns in this gene.

Although introns have no function, a mutation within an intron can have drastic effects. A small change in particular positions can mislead the machinery that edits out the introns, and the wrong protein or no protein at all is manufactured. There appears to be a general mechanism to mark the end of a coding sequence and the start of an intervening sequence: introns always seem to begin with the bases GT and to end with AG. But if there should be a mutation in these pairs then the intron will not be spliced out. Several types of thalassaemia result from the destruction of an exon/intron boundary, and the resulting non-production or abnormality of one of the globins.

Apart from introns, there are many other extraneous and

apparently functionless sequences in human DNA. One might think that the amount of DNA in a genome would be related to the biological complexity of the organism. A bacterium has only about 4.5 million base pairs in its genome whereas a single copy of the human genome has around 3 billion. But a mouse has just as much if not more base pairs in its DNA as a human, and the salamander has more than 30 times as much. Salamander DNA contains around 90 billion base pairs. Plants can have large genomes: maize (corn in the USA) has around 15 billion base pairs and the lily has about the same number as the salamander, 90 billion.

Armed with the knowledge that three bases comprise a codon and that codons correspond to amino acids, it is possible to work out from the size of a protein molecule roughly how many bases were necessary for the cell to fabricate it. The average size of a protein molecule leads to the prediction that, in humans, the average gene ought to be about 1,000 base pairs long. Since a human being is estimated to have no more than 100,000 active genes there ought to be at most about 100 million base pairs in the human genome. In fact, there are 30 times as many. It seems as if no more than about 3 to 5 per cent of the human genome actually contains instructions for the manufacture of proteins. Some of the rest has a role in regulating the activity of genes: switching them on and off, while other portions are important in organizing the structure of DNA and for replication. But the vast bulk of human DNA – something like 90 per cent – seems to have no function at all.

Thus, by the end of the 1970s researchers faced a paradox: the decade had opened up the technologies of DNA analysis which made it possible to think of teasing out the exact sequence of all the base pairs in the human genome. But it looked as though much of that sequence would have nothing to tell them. Would it be worth the effort?

The neat picture painted in previous chapters of the human genome as a library holding between 50,000 and 100,000 books needs to be revised. As already noted, it contains two editions of every book, and physically the library is divided up into forty-six separate bays, corresponding to the chromosomes. But the biggest surprise perhaps is that there are about thirty times as many books in the library as might have been expected, most of them completely meaningless. And it appears as if the genes cannot really be represented by books at all, they seem to be more like fragments of text, not even amounting to separate chapters, scattered virtually at random within much larger books – short stretches of lucidity within a desert of unmeaning.

This more realistic picture of the human genome shows just how enormous are the difficulties confronting those who hunt for specific genes implicated in disease and just how enormous is the task facing those who wish to map and sequence the entire human genome. The ultimate aim of the Human Genome Project is to read every letter of every word in the library that is the human genome – including the vast volumes of nonsense – to obtain the complete set of instructions on how to put together a human being. A different way of gauging the magnitude of the enterprise is to consider the Human Genome Project as akin to conducting a census of every human being living on earth today. Biologists are confident not only that they can conduct that census, but that they can do the equivalent of writing down every individual's full address, and their relationships with other members of their family.

5

The Human Genome Project

NANCY Wexler works from a small office in the Psychiatric Institute attached to Columbia University's Presbyterian Medical Centre, located at the unfashionable end of Manhattan. The walls of the corridor outside her room are covered with extended family trees, detailing the lineage, descent and relationships of fisherfolk from Lake Maracaibo in Venzuela. Every March, Dr Wexler flies to Venezuela for five or six weeks, to collect more data. She now has information on more than 11,000 people included in the pedigrees she has been compiling since 1979: 9,000 of them are still alive, the majority of them under forty. About 650 of the children are pre-symptomatic carriers of the genetic defect which will lead to Huntington's chorea.

The villages around Lake Maracaibo have the highest known incidence of Huntington's in the world and the information gathered from these people about their family ties, coupled with the analysis of their genes, led to the development of a test that can detect the Huntington's gene long before any symptoms are apparent. Nancy Wexler's interest in the research is more than that of disinterested scientist: her mother died of Huntington's. To the distress and anguish of witnessing her mother's illness was added the knowledge that she herself was therefore at risk of having inherited the defective gene that causes the illness.

Huntington's chorea is an invariably fatal degeneration of the central nervous system. It is caused by a defect in a single gene. Huntington's chorea occurs even when the sufferer inherits only one copy of the Huntington's gene. The genetic defect that causes Huntington's is 'dominant' and anyone who is a carrier is also a sufferer. One well-known victim of this disease was the American folk singer Woody Guthrie, who inherited the disease from his mother who had also died of the condition. Huntington's chorea is an agonizing disease in which the victim jerks in involuntary movements brought on by progressive degeneration of the nerves; it leads eventually to dementia and death. It kills, but does so slowly: Guthrie entered hospital in 1954, but did not die until 1967. There is a further cruelty to this disease. It does not set in until the victim is in his or her late thirties or even later, by which time the sufferer may already have had children, each of which stands a 50:50 chance of inheriting the disease.

Classical genetics had made little impact on Huntington's apart from identifying it as a dominant single-gene disease. It had not proved possible to assign the gene responsible to a particular chromosome and no one had any idea of what was actually going wrong biochemically in the body. But new genetic techniques developed in the 1970s have eventually led to a chromosome assignment for the Huntington's disease gene and a DNA-based diagnostic test for its presence, although at the time of writing neither the gene itself nor the underlying biochemical defect have yet been identified.

The techniques used to probe the molecular genetics of Huntington's are those on which the Human Genome Project's attempt to map the whole genome are founded. One of these is the jaw-breakingly named restriction fragment length polymorphism, called RFLP for short.

By 1978 and 1979, studies had shown that when a restriction enzyme is applied to the DNA from different people the DNA fragments produced will usually differ in length from one individual to another. The same fragments of the same stretch of DNA could be twice as long or more in one individual when compared with another. There are two explanations for this variation. One is that at a point in the DNA where one person has a recognition site for the enzyme to cut, others have a mutation in their DNA which alters the base sequence and makes it unrecognizable to the restriction enzyme. The other is that some people have longer stretches of DNA between two restriction enzyme cutting sites than others. This is because of the presence of variable numbers of short repeated DNA sequences. It turns out that human DNA varies from person to person such that, in areas which are not genes, about one in every 500 base pairs or so differs. This variation, known as 'polymorphism', is harmless in these cases, because it occurs outside the genes themselves, but is inherited just like a gene. The consequence is an easily detectable polymorphism in the length of restriction enzyme fragments and these restriction fragment length polymorphisms (RFLPs) are the key to the new era of genetic analysis.

In the late 1970s, the British geneticist Sir Walter Bodmer and his American colleague then working in England, Ellen Solomon, were thinking about RFLPs in relation to inherited disorders of haemoglobin – the haemoglobinopathies (e.g. sickle-cell anaemia). In 1979, they published a paper in *The Lancet*, in which they outlined a way in which the patterns of fragment length could be used as markers to flag the presence or absence of particular gene variants. Since humans inherit two copies of each chromosome, one from their mother and one from their father, a probe for a particular stretch of DNA will pick out two copies from a person's DNA – the maternal and paternal copy. Often these fragments will be considerably different in length, indicating the presence of an RFLP. If the geneticist can get access to the parental DNA as well they can find out which parent carried which variant. The pattern of restriction fragment length acts as a characteristic fingerprint for that particular piece of DNA, allowing geneticists to track its inheritance down the generations. If it turns out that the presence of the RFLP in the DNA is invariably associated with the presence of a defective gene, as deduced from the genetic pedigree, then the RFLP can be used to detect the presence of the aberrant gene even though the gene's exact location, identity and nature is still unknown.

The same thought had occurred to researchers in the USA and in July 1980 a more extensive working-through of the idea appeared in the *American Journal of Human Genetics*. David Botstein, of the Massachusetts Institute of Technology, and Ronald Davis, of Stanford University in California, are said to have hit on the idea together in 1978 while attending a seminar on genetics in the mountains near Salt Lake City, Utah. The seminar had been organized by Mark Skolnick, a population geneticist at the University of Utah. Human genetics was then still very much a business of studying large numbers of closely related people and seeing if it was possible to track the inheritance of a given characteristic through several generations of one family. The extensive genealogical records that the Mormon church in Utah has kept, coupled with the Mormon tendency to have large families, represented a goldmine of data for human geneticists.

The detailed experimental work to prove that the idea of RFLPs as genetic markers was feasible was done by Ray White at the University of Massachusetts Medical School. The four Americans realized from the first that at last they could easily detect landmarks throughout the human genome and so had a way of mapping human genes in relation to these landmarks. The crucial point with RFLPs is that the markers do not need to be inside a gene, they need only be near enough to it on the chromosome for marker and gene to be almost always passed on together from parent to child.

Here at last was the tool by which researchers could look at variation between individuals, not by probing some obvious physical characteristic or biochemical trait (the phenotype) which might not fully reflect the actual genetic constitution (the genotype) nor even by looking for the proteins expressed by the genes, but by analysing DNA itself directly.

These new techniques of genetic analysis, together with the co-operation of the Venezuelan families, made it possible to think of tackling the Huntington's gene for the first time. Wexler and her colleagues persuaded the fisherfolk of Lake Maracaibo to donate small blood samples which were flown back to the USA to the laboratory of James Gusella at the Massachusetts General Hospital in Boston. Nancy Wexler and her geneticist colleagues tracked the inheritance of Huntington's back through the extended Venezuelan families and, in Boston, the molecular biologists began to search through the DNA from the blood samples for a characteristic pattern of restriction fragment lengths present in the DNA of those people who were already known to be suffering from the disease, but absent from those definitely free of the disease. In this way, they hoped to identify genetic markers that were inherited in common with the aberrant gene, but which were absent from the DNA of those who carried the normal variant. In 1983, they found such a pattern and eventually linked it to chromosome number 4. Some people who had not yet shown any symptoms of the disease also exhibited the charac-teristic RFLP – and these later came down with the disease. The technique had proved its predictive power and the gene was assigned to chromosome 4. It was the first triumph of 'reverse genetics' – of looking for patterns in human DNA that might reveal future disease long before the symptoms themselves became apparent.

At last techniques were available to begin the Herculean task of mapping the positions of genes and genetic markers in the human genome and then deciphering the sequence of genetic instructions written into the fabric of DNA itself.

The first initiatives in the Human Genome Project were made in the USA, which has since developed the best financed and co-ordinated programme. For that reason therefore, the genesis of the US Human Genome Project will be described in some detail. In 1984, the molecular biologist Robert Sinsheimer put forward the idea of establishing an Institute to Sequence the Human Genome at the University of California at Santa Cruz, of which he was chancellor. It was to be a prestige project, the idea for which had come out of a multimillion dollar project to build an astronomical telescope. In 1984, the University of California received $36 million to construct a 10-metre telescope at the Lick observatory. Nor was this the only 'Big Science' project being considered at the time. Physicists studying the fundamental constituents of matter – the elementary particles – were beginning a campaign to secure funds to build a huge particle accelerator, the superconducting supercollider or SSC, the cost of which was reckoned in billions of dollars. Many states and universities were keen to attract such a lucrative project to their territory – and Professor Sinsheimer was a member of the Californian team. In the end, the SSC went to Texas, and the large optical telescope came to nothing, but the idea that biology could be Big Science was planted in Sinsheimer's imagination. 'It was thus evident to me that physicists and astronomers were not hesitant to ask for large sums of money to support programmes they believed to be essential to advance their science,' he later told the British magazine *New Scientist*. In addition, an Institute to Sequence the Human Genome would have the added attraction of bringing Santa Cruz up to the same level of academic standing as the better-known campuses of the University of California at Berkeley and Los Angeles.

In May 1985, Sinsheimer called a meeting of about a dozen of the USA's top molecular biologists to discuss how such a goal could be achieved. According to Norton Zinder, professor of molecular genetics at Rockefeller University, New York, who was later to chair an advisory committee to the Human Genome Project:

Initially, almost all the scientists at the 1985 meeting were highly skeptical. Two almost diametrically opposed positions were formulated. One was that we would not learn enough to make the large and expensive effort worthwhile. This feeling was in part based on the fact that about 90 per cent of the human genome may have no function ... A related idea held that a large applied-science programme would distort the workings of science. On the other side were those who felt that we would learn far too much. Having the human genome at hand might provide an infinite number of new reasons for genetic discrimination by employers and insurance companies; it might even inspire Nazi-like eugenic measures. At a minimum the number of genes known would increase manyfold and there would still be a long lag between the time a disease gene is identified and the time treatment might be available.

The issues that Zinder identified in his essay in *Scientific American* of July 1990 still dominate discussions of the Human Genome Project. In the end, the idea of a Genome Institute at Santa Cruz was never realized, but the idea that there ought to be some co-ordinated effort to map and sequence human genes began to gather momentum. One enthusiast was the Nobel Prize-winning biologist Renato Dulbecco. In the autumn of 1985, he advocated sequencing the human genome during a speech at the Cold Spring Harbor Laboratory in New York. This caught the attention of the director of the laboratory, James Watson.

Independently of Sinsheimer's efforts at Santa Cruz, the US Department of Energy (DOE) started to get in on the act. This may seem slightly bizarre until it is realized that the Department has had a long-standing interest in human genetics and in mutation because of its military and civilian nuclear programmes. At the end of the Second World War, the US Government and Congress decided not to entrust the production of the United States' nuclear weapons to the military men in the Pentagon, the US Department of Defense, but to assign it to a civilian agency, the US Atomic Energy Commission. The Commission had responsibility not only for the design and production of nuclear weapons but also for the development of civil nuclear power. In this latter role, it had responsibility both for promoting civil nuclear power and for regulating the safety of nuclear power stations. In the mid-1970s, it was felt that these two roles conflicted and so the Commission was split. A Nuclear Regulatory Commission was set up to oversee safety and the new Energy Research and Development Administration (ERDA) was given a broad mandate to investigate other forms of energy than nuclear power. The ERDA quickly metamorphosed into the Department of

Energy, which retained responsibility for nuclear weapons design and production as well as for the safety of the reactors and other processes used in weapons production. For most of the post-war period, therefore, the DOE and its predecessors have had an interest in human genetics because of the need to understand the effects of radiation on human beings and their genes. One of the principal techniques used in this work is visual examination of chromosomes for any radiation-induced abnormalities. In 1983, the two main nuclear weapons laboratories, at Los Alamos and Lawrence Livermore, started work on a Gene Library Project. The laboratories had become involved in new techniques for separating and sorting chromosomes – particularly a technique known as 'flow cytogenetic analysis'. In this, chromosomes are mixed with fluorescent markers. Because different chromosomes bind different amounts of the markers, it is possible to sort them one from another by shining laser beams at the marked chromosome and measuring how much marker each has taken up. By 1986, this approach could successfully sort all the chromosomes except 10 and 11. And by February 1986, the national laboratories had also established a library of human DNA fragments.

The Department of Energy has an Office of Health and Environmental Research (OHER) to oversee its radiation safety work. In 1986, Charles DeLisi, head of OHER, started promoting the idea that the DOE should take a greater role in modern approaches to human genetics through the new molecular biology. He recognized that sequencing the human genome was going to be a big project, and suggested that the DOE, with two major nuclear weapons laboratories in its portfolio, was well used to managing big scientific projects. The timing of these suggestions was perhaps surprising: the Cold War had become more intense around this time and the United States was committed to building and testing more nuclear weapons; in addition, President Reagan was promoting the idea of a 'Star Wars' defence against nuclear attack which would involve lots of work for the laboratories, developing new kinds of nuclear weapons.

But in the event, the Strategic Defense Initiative came to very little and events in Eastern Europe removed much of the political rationale for an extensive nuclear weapons programme. Even without the collapse of the Soviet Union, it is unlikely that the Reaganite agenda could have been sustained. In the 1990s, the weapons laboratories are engaged in energetic self-preservation by diversifying out of weapons-related work. It may be that DeLisi recognized subliminally that, despite appearances in the early 1980s, the prosperous times for nuclear science and technology were drawing to a close, and that now was the appropriate moment to switch into molecular biology. The DOE organized a major scientific gathering in Santa Fe, New Mexico, in March 1986 to discuss DeLisi's ideas. The meeting enthusiastically endorsed the idea of sequencing the human genome, and the DOE picked up the initiative. Although sequencing the human genome is an expensive enterprise, the costs of the early stages are relatively modest and, with its easy access to large amounts of money, the DOE forged ahead with human genetics.

Thus, by mid-1986, there was a complicated situation in the United States. Although the best biological information would come from a map of the human genome, most emphasis had been put on the much more laborious and expensive business of sequencing. In addition, there were two separate communities of scientists involved. On the one hand, the molecular biologists in the universities and other life-sciences research institutes would naturally look to the NIH, which channels most federal funds for biomedical research, but the NIH seemed hesitant and unwilling to embark on such a huge commitment. On the other hand, the Department of Energy had large, well-equipped and well-financed national laboratories, which looked as if they could afford the Human Genome Project out of the small change from their nuclear weapons design and development budgets.

Some very eminent molecular biologists opposed the whole idea of sequencing the human genome. More were suspicious of allowing the DOE to become the lead agency in the project. Throughout 1986 and 1987, the debate became heated. The Whitehead Institute in Massachusetts, where much of the work on mapping the genome had been done, became a focus of opposition to the sequencing project. David Baltimore, a Nobel Prize-winner and then director of the Whitehead, wrote in *Science*: 'The idea is gaining momentum. I shiver at the thought.' David Botstein, also of the Whitehead, who had been one of the co-authors of the original paper in 1980 which set out the basis of how to construct a map of the human genome, added his voice to the doubters. Some three months after the DOE's meeting in Santa Fe, in June 1986, the Cold Spring Harbor Laboratory organized a symposium on the 'Molecular Biology of *Homo sapiens*'. Botstein told the symposium that starting to sequence the human genome then would

79

be 'like Lewis and Clark going to the Pacific one millimetre at a time. If they had done that, they would still be looking.' Another Whitehead Institute scientist, Robert Weinberg, told *New Scientist*, 'I'm surprised consenting adults have been caught in public talking about it. It makes no sense.' But such opposition was not confined to the east coast. James Walsh from the University of Arizona and Jon Marks from the University of California at Davis wrote in *Nature*, 'Sequencing the human genome would be about as useful as translating the complete works of Shakespeare into cuneiform, but not quite as feasible or easy to interpret.'

The scientists' concerns were about the size and cost of the effort that would be involved in sequencing the human genome. Few disputed that, ultimately, mapping and sequencing the human genome would be a major advance for the life sciences, but many were worried about the best route to the desired destination. In an editorial in the American journal *Science*, the Editor, Daniel Koshland, put the case for a genome project very simply: 'The main reason that research in other species is so strongly supported by Congress is its applicability to human beings. Therefore the obvious answer as to whether the human genome should be sequenced is "Yes. Why do you ask?".' But there was no tradition in biology of the sort of big science to which physicists and astronomers had become accustomed. Watson and Crick had solved the structure of DNA while working from a tiny office at the Cavendish laboratory in Cambridge; and in the late 1950s, Crick even had to work out of a converted bicycle shed in the yard at the back of the laboratory.

Although instruments and apparatus had got more sophisticated (and correspondingly more expensive) over the years, the character of molecular biology remained that of a cottage-industry type of science. Large institutes consisted of small groups each doing its own work in its own laboratories – the attraction of working in a large unit was the conversation and discussions with other biologists, the sharing of ideas, not a quasi-industrial division of labour. Biology had seen no big enterprises, such as the Manhattan Project which brought thousands of scientists and engineers together to build the atomic bomb during the Second World War, nor was it accustomed to the division of scientific labour practised by astronomers and physicists whereby a research student, for example, might play the most minor part in a co-operative enterprise, building a detector perhaps for someone else to use in an experiment. Molecular biology was still very much a science for individuals and many practitioners feared to lose the style of science to which they had become accustomed. For there was little doubt in anyone's mind but that the task of sequencing the human genome would require a huge amount of work, much of it boring and repetitive. While the mechanical labour of getting the sequence was being done, who would have time or energy left over to interpret the biological significance of the results?

A second more mundane concern was money. Sequencing the human genome would be expensive, but the funding for the project was likely to come from the same pot as the money for the rest of molecular biology. Walter Gilbert, the physicist turned molecular biologist who invented one of the ways of sequencing DNA, estimated the price at around $1 a base. This would mean spending $3 billion over perhaps 15 years to sequence all 3 billion base pairs of the human genome. The sequencing project therefore would need separate funding, or existing research – which often had a more immediate biological pay-off – would suffer in the competition for funds. *Science* ran an editorial suggesting that the DOE ought indeed to run the whole venture, to avoid draining money from the NIH and the scientists supported by it.

But by 1986, James Watson had become convinced that an attempt on the human genome was desirable and feasible. He was also convinced that the project could not be left to the bureaucratic organization of the DOE, but that it had to be led by scientists and driven by the perceived needs of the science. This meant that the NIH had to be involved. He rejected the arguments put forward by the DOE that it should be the lead organization in the project as 'forced and disingenuous.'

At this stage, discussion and debate had reached such a peak that several separate investigations were mounted into the issue. One politically powerful initiative came when the US Congress's Office of Technology Assessment set up a task force to examine what was going on. Other reviews were carried out by the DOE itself and by the Howard Hughes Medical Institute. But the crucial step came when the National Research Council decided to investigate the issue. The Council is the operating arm of the US National Academies of Science and of Engineering, and as such, speaks for the scientific community with more authority than perhaps any other organization within the USA. It held a meeting at Woods Hole, Massachusetts, in August 1986 and then set up a high-powered committee to examine the arguments. The Academy study turned the project on its head, emphasizing the benefits of the genetic atlas – of mapping the human genome before starting to sequence the base pairs. It reiterated the previous informal estimates of the cost of the enterprise, suggesting that annual funding of about $200 million over 15 years would be required for success. The report also stressed the importance of studying the genomes of organisms other than humans, in order to be able to make biological sense of the human data. For obvious reasons of ethics, it is impossible to do experiments in human genetics. No one can think of modifying a couple of base pairs in human DNA deliberately, just to see what happens. Less moral difficulty attaches to experiments in the genetics of, say, bacteria. The NRC committee stated that:

To derive the major benefits from a human genome sequence, it will be necessary to have an extensive data base of DNA sequences from the mouse (whose genome is the same size as that of the human) and from simpler organisms with much smaller genomes, such as bacteria, yeast, *Drosophila melanogaster* (a fruit fly) and *Caenorhabditis elegans* (a nematode worm) . . . To succeed, therefore, this project must not be restricted to the human genome; rather it must include an extensive sequence analysis of the genomes of selected other species.

Such an analysis meant more than doubling the amount of work to be done. Watson claimed the credit for widening the scope of the project. It was 'My most important contribution to the project as a whole,' he later told *New Scientist*.

While the blue-ribbon panel of scientists was finalizing its report, federal money was voted to support human genome research by the NIH. In December 1987, $17.2 million was earmarked for the NIH work. In parallel the DOE had received about $12 million for its research on the human genome. By then the department had lost its most visible champion of genome work – Charles DeLisi had left his job as director of health and environmental research in 1987 – and its standing in the genome project was to diminish progressively thereafter.

The National Research Council and the Office of Technology Assessment reports were both published in 1988. In February 1988, the then head of the NIH, James Wyngaarden, called another meeting. This meeting, in Reston, Virginia, was chaired by David Baltimore, who had previously been critical of the whole idea of sequencing the genome. But the work of the NRC had converted the basic thrust of the programme towards mapping rather than sequencing and towards biological understanding (through study of other species' genomes) rather than the brute-force acquisition of data. The Reston meeting transformed the attitude of the NIH and set out detailed goals that the programme should achieve.

Early in May 1988, Wyngaarden invited Watson to take on the responsibility of running NIH human genome research. Watson took the job because, as he later wrote, 'I realized that only once would I have the opportunity to let my scientific life encompass the path from double helix to the 3 billion steps of the human genome'. His appointment, according to Norton Zinder, brought about 'a quantum jump in the program's credibility'. On 1 October 1988, Watson was named Associate Director for Human Genome Research at the National Institutes of Health, with a 1988–89 budget of more than $28.2 million – some $10 million more than the DOE's genome research budget for that year. On the same day, the NIH and the DOE signed a Memorandum of Understanding on how the two agencies would co-operate over genome research. The US Human Genome Project was now properly launched – and with the NIH, rather than the DOE, clearly at the helm.

On his appointment in 1988, Watson's programme at first came under the aegis of the National Institute for General Medical Sciences whose head, Ruth Kirschstein, had previously expressed opposition to the idea of a targeted programme to analyse the human genome. The Office of Genome Research started with a staff of two, plus Watson. Within a year, however, it had expanded into the National Center for Human Genome Research, independent of the National Institute for General Medical Science, and had command of its own budget. By October 1989 this had risen to nearly $60 million – more than double the DOE allocation of $28 million. The Center's staff numbers grew to thirty and its budget to around $108 million by 1991. The DOE's funding, however, was growing much less swiftly and amounted to only about $46 million over the same year.

Watson was characteristically forthright in his advocacy of the project and of the leading role that scientists rather than administrators or bureaucrats must take in its development. He caused consternation in 1988 when he suggested that the most effective way of setting about the project would be to share out the chromosomes among different countries and different laboratories. He was reported as saying at one press conference that, because chromosome 1 was the largest (and by implication would require the most work for least return) 'we'll give that to the Soviet Union'. He cause more international flutterings when he suggested that scientific data obtained in the course of the US project could be denied to Japanese scientists if the Government of Japan did not support a national programme of research into the human genome of the same proportions as that in the USA. Two years later, in an article in *Science*, Watson was more circumspect, but there was still a hint of iron beneath the velvet prose. He wrote that 'The idea that the various human chromosomes will be divided among the various laboratories is far from today's conventional wisdom.' But, he pointed out, 'extensive multiple overall mapping efforts only exist for chromosome 21, the smallest human chromosome, which contains among others a gene that leads to increased susceptibility to Alzheimer's disease ... To my knowledge, the number of investigators wanting to make complete, high-resolution physical maps of a specific chromosome is less than ten. Our real problem may be persuading capable teams to focus on those chromosomes that still have no champions.' In other words, there would still be some pressure to persuade people to get into the business of mapping complete chromosomes. The international aspects of the project were similarly couched in more diplomatic, but still pointed language:

Early sharing of the human DNA database is much more likely to occur if large-scale mapping and sequencing efforts are undertaken by all those major industrial nations that will want to use this data. It is too early to ask what we should do if we identify one or more countries that have the economic clout to join in the effort but that apparently do not intend to do so, hoping instead to take advantage of the information once it becomes publicly available. I do not like to even contemplate such a possibility, since Congress and the public are likely to respond by wanting to move us toward a more nationalistic approach to science. This alternative is counter to the traditions that have allowed me to admire and enjoy the scientific life. The nations of the world must see that the human genome belongs to the world's people, as opposed to its nations.

But the sequence itself was and is still a long way off. After the initial flush of enthusiasm of the mid-1980s, it had become apparent that the available techniques were too slow and too expensive to make it worth starting immediately to sequence all 3 billion base pairs in the human genome. In the summer of 1989, at a retreat at Cold Spring Harbor, representatives of the NIH, DOE and other invited experts drew up a five-year programme for the US Human Genome Project. By then the cost of DNA sequencing stood at between $2 and $5 a base pair and they ruled out large-scale sequencing of the human genome until the true cost of sequencing had fallen to no more than 50 cents per base pair. They also noted that it might not be enough simply to develop and refine existing methods of DNA sequencing, 'Therefore, entirely new approaches to DNA sequencing will also be encouraged.' The joint report which stemmed from that meeting, 'Understanding Our Genetic Inheritance – The US Human Genome Project: The First Five Years', went to the US Congress in February 1990. It updated the National Research Council and Office of Technology Assessment reports of a couple of years previously and set specific goals for the research to be done by 1995. The sequence of the human genome will now certainly not be available until more than half a century after the elucidation of the structure of DNA.

Like any explorers venturing into unknown territory, the scientists who wished to embark on the great adventure of sequencing the human genome decided first to equip themselves with maps. The Human Genome Project will divide into two stages, the first of which is to compile a map of the human genome. In fact, two maps will be compiled, reflecting the dualism between the genes and the chemistry of DNA. One map will be a genetic map, relating known genes or genetic landmarks to each other; whereas the other map will be a physical map, relating known sequences of DNA to each other. The first goal set for the US programme by that meeting in Cold Spring Harbor was to complete a full genetic map with markers spaced about 2 to 5 centimorgans apart (see Chapter 4). There is not an exact equivalence between genetic distance measured in centimorgans and physical distance measured in base pairs, but on average a centimorgan represents around 1 million base pairs. The members of the panel estimated that between 600 and 1,500 different markers would be needed to achieve such a genetic map – eventually, for a complete map of the human genome, about 3,000 well-spaced markers will be needed.

It might seem at first sight that mapping is a mere preliminary which will yield much less information than the sequence itself. But this is a false perception. The importance of the map had been recognized very early on and had merely been obscured by the mid-decade enthusiasm for sequencing. According to a talk given by Samuel Ajl, vice-president of the US March of Dimes Birth Defects Foundation, in November 1979, 'The gene map is the inner picture of the species, its immortal part. The gene map may tell us not just where the genes are but why they are there.' More recently, the contemporary historian of modern biology, Horace Freeland Judson, has written that if the sequence was, so to speak, the terrain to be mapped,

the sequence by itself is dry and uninformative – an all-but-endless string of A, T, G, C, in no predictable order. But maps are never the territory mapped. They are at once less than the territory and more. They are abstractions from the territory, leaving much out; and they are labelled, the features of the terrain identified. Every natural territory generates many different maps and the differences spring from the purposes to which the maps are to be put. The political map differs from the road map which differs from the map of natural resources as seen from mapping satellites and so on. Yet all maps share one general characteristic: the abstracting and the labelling establish a set of relationships. In short, by omitting much of the territory, maps gain explanatory power. By being less than the sequence, the maps of the genome will be far more.

The second goal was to construct a purely physical map of the genome. This essentially involves setting up the fragments of human DNA contained in a 'library' in their correct order – arranging the fragments in the order that they appear in the chromosome itself. These overlapping sets of clones of DNA are known as 'contigs'. The idea was to construct a map in this way with markers spaced at intervals of around 100,000 base pairs.

The US researchers also emphasized the need to refine the technology of DNA sequencing to bring the price down from the present $5 a base pair to around 50 cents. They envisaged perfecting existing techniques and applying them to the sequencing of continuous stretches of DNA of up to about 10 million base pairs in aggregate.

These plans for the US project also paid full attention to the mouse genome and to the other model organisms. Other goals included developing the computer databases and software capable of handling the flood of data that will eventually be generated by mapping and sequencing efforts, training more graduate and post-doctoral research workers to contribute to the work, and to encourage close working relationships with industry and facilitate technology transfer from academia to the medical community and to commerce.

The US Human Genome programme is far and away the most lavishly financed and is generating the lion's share of publications in the international scientific journals. In second place, both in terms of financing and scientific productivity, lay the British genome programme, which was conceived in much less public circumstances than the American one, a feature characteristic of the cultural difference between the two English-speaking countries. Senior officials and scientists working for the UK Medical Research Council (MRC) watched with growing unease the development of new genetic techniques in the USA and became worried that British scientists (and, further down the line, perhaps the British pharmaceutical industry) would fall behind and either be denied the fruits of such work or have to pay costly licence fees in order to get access to it. Among the techniques which they saw being developed were yeast artificial chromosomes and pulsed-field gel electrophoresis.

One of the dominant figures in British molecular biology, the South African-born Sydney Brenner, convinced the Council that it should seek more money from the government to set up a special programme of research into the human genome. At this time, around 1987/8, money for basic scientific research was very tight, but nonetheless a bid for more money was made to the standing committee of senior scientists, known as the Advisory Board for the Research Councils, which determines the distribution of whatever money the government does allocate to science. The Advisory Board recommended that funds be allocated to the MRC from the science budget and earmarked for human genome research. At about the same time, the idea came to the attention of the then Prime Minister, Margaret Thatcher, and caught her imagination.

The British Government also receives advice from another small committee, known as the Advisory Council on Science and

Technology (ACOST), which consists of the 'great and the good' from both pure science and scientifically based industry, and which the Prime Minister herself occasionally chaired. Despite this level of interest, ACOST, which meets and reports in complete secrecy, has been largely ineffectual and from what little is known of its advice to Government, most of it appears to have been comprehensively ignored. However, it appears to have been successful in this instance in backing the decision, already taken by the Advisory Board for the Research Councils, to establish the British genome project. In 1988, one of ACOST's members, Keith Peters, Professor of Medicine at the University of Cambridge, who had been working closely with Brenner, put up the idea of a special programme on the human genome to ACOST. With Prime Ministerial backing, the MRC's directed programme of research was quickly secure and, for a few years at least, virtually immune from competing demands on the science budget. The amount of money allocated under the pro-gramme seems modest – about £5.9 million in 1992 – but is sup-plemented by other grants from within the council, bringing the total up to about £20 million.

The British decided that, with their modest resources, they should not attempt to duplicate the American approach, for they would inevitably come second in any direct race. Instead they decided to concentrate on analysing the DNA that is active – DNA that is actually expressed and turned into protein. As noted in Chapter 3 some 90 per cent of the human genome seems to play no part in the specification of proteins. Rather than sorting through all this 'junk' DNA, the British, at Brenner's instigation, decided that it would be more cost-effective to turn to what is known as complementary DNA or cDNA. When a cell makes a protein, it first transcribes the relevant gene into messenger RNA which carries the genetic information out of the nucleus to the cell's protein manufacturing machinery in the cytoplasm. The RNA is spliced to remove introns, the intervening sequences of 'junk' information, before protein synthesis commences. Brenner's idea was to pull spliced messenger RNA out of the cell and use the enzyme reverse transcriptase to make a complementary copy of the sequence in cDNA. In this way researchers would obtain only the 'useful' sequence – that which encodes protein.

There are two problems with this approach. One is the sheer technical difficulty of sorting through the thousands of messenger

RNAs within a cell to pick out one particular sort. Moreover, the one that you want may be present in only small amounts compared with other more abundant mRNA molecules. The second problem is of a slightly different sort. No one really knows whether some of the 'junk' DNA may not, after all, be fulfilling some biological purpose; it is really only a matter of conventional wisdom rather than deep biological insight that dismisses 90 per cent of the human genome as being just so much junk.

There are certainly indications that introns – the intervening sequences of DNA which are not transcribed into messenger RNA and therefore do not contribute to protein synthesis – may have played a role in evolution. Introns spread a gene over a greater length of DNA than is the case in analogous bacterial genes and thus increase the probability of recombination between one gene and another when two chromosomes cross over. In this way, they increase the rate at which an organism could 'invent' new proteins but, because the point of recombination will usually be in the middle of an intron, the probability of a damagingly abberant protein structure is reduced. The genes within human DNA which specify haemoglobin contain what may be an example of evolution in action. Within the cluster of genes which contain the recipes for the different haemoglobins used in the course of development of a human embryo and foetus are 'pseudogene' sequences which look very similar to the globin genes but which do not give rise to mRNA nor, thus, to protein. These sequences may be former genes which have been mutated so that they no longer function, or they may represent regions undergoing evolution to some new function – it is not known which is the case. It could be that this material does play other important but hitherto unsuspected roles which will only become apparent when the junk DNA has been sequenced.

Concentrating solely on cDNA could risk missing something very important. The British therefore have a second string to their bow: to sequence the entire genome, not of something as complicated as a human being, but of a simpler organism which will act as a 'model' for the more complex case. If the junk DNA does have some purpose, perhaps this will become clear from sequencing a simpler creature. In addition, the technologies which have to be developed to carry out the sequencing of even a simple organism can be carried over to more complex tasks, such as the human genome. The model organism chosen by the British is the microscopic nematode worm *Caenorhabditis elegans*. More than twenty years ago, Sydney Brenner suggested that this might be an ideal creature in which to study development from egg to adult animal. The worm is small – about one millimetre long – and has a very simple body plan. It reproduces about once every three days – always an advantage for researchers trying to study the effects of mutations – and it replicates prolifically, producing about 300 offspring per parent. But best of all, the worm is almost completely transparent, which means that the growth of its internal tissues can be seen through the optical microscope. Over the past couple of decades, Dr John Sulston at the Laboratory of Molecular Biology at Cambridge has led the research which has tracked the development and lineage of every one of the nematode's 945 cells, making it one of the most intensively studied organisms in the world. It was natural therefore to turn from understanding the nematode at the cellular level to analysing it at the level of its genes. This project is a collaborative one with the American Human Genome Project. The work is being pursued jointly by Sulston at the Laboratory of Molecular Biology in Cambridge and by Dr Bob Waterston at Washington University, St Louis. At present rates of sequencing, however, it will be nearly a decade before even the worm's genome is completely sequenced.

France has two institutions which are fast eclipsing British efforts in the human genome initiative. In 1984, the Centre d'Etude du Polymorphisme (CEPH) was set up in collaboration with the United States, supported in part by private funds and by the French Ministry for Research and Technology. Under the direction of the Nobel Prizewinner Jean Dausset, CEPH collects reference samples of cultured cells taken from members of extended human families. Most of the current human genetic map has been constructed using the resource supplied by CEPH. More than sixty large families now comprise the reference panel, coming from France, Pennsylvania, Utah and Venezuela. In 1990, CEPH's total budget was around FF40 million (roughly £4 million). A second large project in France is Genethon, which opened its new laboratories at Evry outside Paris towards the end of 1990. This is funded by the Muscular Dystrophy Foundation out of the money it raises from annual television appeals – the one in 1990 gathering more than FF300 million. It is intended to work closely with CEPH to gather more samples from families and to map disease genes. The cDNA approach is favoured throughout Europe, and France too is pursuing this line of research. Government funds earmarked for human genome research are in the region of about FF100 million a year (roughly £10.4 million), but the government also funds general genetics research to the tune of a further FF150 million a year (about £15.6 million). Among the other European countries, Denmark, Italy and Germany have national genome initiatives, although they are not as ambitious or well coordinated as the British and French. Denmark has made an important international contribution, for it, like France, has compiled samples from reference families. There are around 9,000–10,000 individuals represented in the Danish cell bank, but the Danish families are smaller than those in CEPH.

Of all the industrial nations, Japan has been slowest off the mark to set up a co-ordinated programme of research into the human genome. Some of the reasons for this may be cultural and founded in different attitudes towards inherited disease in Japan. Although the implications of the Human Genome Project will ultimately touch everyone and will range much more widely than the diagnosis and treatment of genetic diseases, it is none the less true that much of the impetus in the USA, for example, has come from individuals (like Nancy Wexler) and from support groups who have personal interests in specific inherited conditions. Their existence puts pressure on the medical and scientific establishment to investigate the disease. In Japan, in contrast, there are few such pressure groups, perhaps because more stigma attaches to an inherited disease. In addition, Japanese people tend to have small families, which makes tracing inherited characteristics that much more difficult. Until recently, apart from conditions such as Huntington's where 40 families had been traced for a co-ordinated research project, only local doctors were engaged in research into inherited disease.

Japan does aspire to be a major player in genome research. It has, however, adopted a very cautious approach, spending only comparatively small sums on preparatory studies: to run workshops, to review what research is already being carried out in the country, and to encourage collaboration among researchers. One project, which has been running since 1981, is a technology development project, known as RIKEN, to produce an automated DNA sequencer, but progress has been slow. One of the difficulties in assessing the Japanese genome programme, and it is common to all science in Japan, is that there is a confusing plethora of different government ministries involved in funding different parts of it, and it is not always clear whether there is any co-ordination among them. The Ministry of Education, Science and Culture, the Science and Technology Agency, the Ministry of Health and Welfare, and the Ministry of International Trade and Industry all have some involvement. In addition, this being Japan, there is considerable interest in the genome of the rice plant, where research is being funded by the Ministry of Agriculture, Forestry and Fisheries.

At the very first stirrings of international interest in the genome project – with many nations motivated not least by a reluctance to be left behind the Americans in the biggest project in biology – it became clear that some international forum was needed. In 1988, at a meeting in Cold Spring Harbor, researchers decided to set up the Human Genome Organisation (HUGO) to co-ordinate international efforts and to try and minimize duplication and overlap. It was headed by the American geneticist Victor McKusick, who was succeeded by the British director of the Imperial Cancer Research Fund, Sir Walter Bodmer. Its legal headquarters are in Geneva, but its operational offices are in London, Bethesda and Osaka. It has started work with grants from major charities, such as the Wellcome Foundation Trust in the UK. But HUGO has found it difficult to make a mark and, without any funds of its own with which to finance research, appears doomed to an impotent existence of offering advice which no one need act upon.

The fragility of international co-operation and the influence on science of seemingly irrelevant political events was demonstrated at the time of the 1991 Gulf War. The International Genome Database resides on computers in Baltimore and many European scientists make extensive use of a suite of programs known as 'the Wisconsin Package' for computerized DNA analysis. But for reasons best known to itself, the US Department of Commerce placed an embargo on access to the computer programs by non-US nationals – apparently in the belief that the programs could be of use to someone embarking on a biological warfare project. Although this notion is completely without foundation, the NIH were unable for some time to get the embargo lifted.

Prospects for international co-operation were dealt a further and more severe blow in mid-1991 when it was announced that the NIH was applying for patent protection to cover DNA sequences which its researchers had discovered. The issue was made more piquant by the fact that the sequences being patented are of genes expressed in the brain, so not only would the US Government own part of the human genome, it would also have succeeded in patenting part of the human brain.

A further complication was that no one knew what genes these sequences came from, let alone their function. One of the difficulties with research into the human (or any other genome) is for one researcher to be sure that the restriction fragment upon which they are working is the same as that isolated by another, so that they can compare results. This need for a common reference system has been addressed by a proposal from US researchers, that fragments of the genome should be identified by 'sequence tagged sites'. The idea is simple: someone who isolates what they think is a new fragment of human DNA should sequence part of it – a stretch that is long enough uniquely to identify that fragment. No one, however, imagined that anyone would pull out fragments virtually at random from brain cells or anywhere else and then apply for a patent to cover the identifying sequence.

The issue raised a storm of controversy within the USA, especially as it appeared that the NIH had decided to take this action without telling James Watson, the head of its National Center for Human Genome Research. Watson adamantly opposed the idea, for he realized that it would render impossible the free flow of scientific information and thus would make international collaboration more difficult and hinder progress. Instead of publishing openly in the scientific literature, everyone would hide their own data until they had secured patent protection for it and no one would know what anyone else was doing. Taken to the extreme, this might mean that each country would have to sequence all the human genome itself, in secrecy, hoping that the parts which it had identified would not already be covered by someone else's patent.

The issue became international and political, with protests at government level from Britain and France. The French considered that human genes simply should not be patentable at all. The British, as usual, tried to compromise. Their position was that gene sequences 'of no known utility' (i.e. the random sequences which the

Americans were trying to patent) should not be covered by patents and no one should seek patent protection for them. On the other hand, sequences whose function was known (the mutant cystic fibrosis gene for example) could be patented. The British Government added a rider: while it was opposed to the idea of patenting sequences of no known utility, it would none the less patent everything it had got, if the Americans persisted.

International co-operation was further imperilled in early 1992, when a private American company tried to attract John Sulston from Cambridge to the United States. The single most advanced area of genome sequencing is the nematode worm on which Sulston at Cambridge and Waterston at St Louis are working. They have developed not only sequencing techniques, but also tools for computer data analysis and ways of organizing the work, which will be invaluable for the larger-scale sequencing of the human genome. All these developments have been carried out with public funding. The private company, to be established in Seattle, wanted both Sulston and Waterston to join it and to bring with them all the expertise built up over the years, so that the company could offer itself as a contract gene sequencing enterprise. When the US genome project moved to large-scale sequencing it would find that one organization had a virtual monopoly of the technology, and that that was a private company which the government would have to pay to carry out the work – even though it had already paid for the process by which the company had acquired its expertise. Not only international collaboration, but the health of the US domestic programme was thrown into doubt by the proposal. Once again, Watson was energetic in his opposition, this time successfully. Both Sulston and Waterston decided to stay put.

Both incidents brought James Watson into conflict with other parts of the US Government bureaucracy and in April 1992, the NIH launched an ethics investigation into an alleged conflict of interest between Watson's position as director of the National Center for Human Genome Research and his shareholdings in private biotechnology companies. It was widely believed within the US scientific community that this investigation was merely a pretext, and that there were no grounds whatsoever for thinking that Watson was subject to any conflict of interest, let alone anything improper. He has certainly not maximized his personal wealth by exploiting his position as one of the discoverers of the DNA double helix. Many scientists at the forefront of the genetic engineering revolution in the early 1970s made themselves multimillionaires by investing in start-up biotechnology companies. There is no evidence that Watson followed suit.

Instead, it is believed that he fell victim to a clash of personalities and to internal politics within the NIH. He had, in any case, intended to stay with the genome research effort for only four years or so, in order to see it safely established. In April 1992, he announced that the burden of coping with two jobs, as director of the National Center for Human Genome Research and as director of the Cold Spring Harbor Laboratory, was proving too great. He therefore resigned from the NIH job and returned to Cold Spring Harbor, where he has been director since 1968. Some sense of the internal tensions within the NIH may be gathered both from the abruptness of his decision and from the fact that the director of the NIH accepted his resignation without expressing a word of regret. No provision had been made for an orderly succession: the NIH hurriedly appointed someone from within its own ranks as caretaker director until Francis Collins, the finder of the CF gene, agreed to take on the job.

Watson had provided the US Human Genome Project with charismatic and energetic leadership. His departure took place at a time when the project looked troubled. At the time of writing, the issue of patenting is still unresolved. In addition, there will be increasing interest by private companies in the fruits of the project and in the business of sequencing. The human genome is so long and complicated that to sequence it will require a quasi-industrial organization, not the type of creative research which one finds in the typical molecular biology laboratory. But involvement by commercial companies will provide fertile ground for potential conflicts of interest over the development of technology and the ownership of the results. These issues will grow rather than diminish. Watson's successor is faced with the task not only of providing leadership to the scientific community, but also of developing a new relationship with private enterprise.

It is clear that after only a few years of operation, the US Human Genome Project has stumbled heavily over one of the first social and moral issues to be raised: who owns the human genome? Who will have rights of access to the fruits of the research? So far, the official American response – as evinced by the NIH – seems to be that the

95

human genome, the innermost essence of humanity, is just another resource for commercial exploitation. The danger against which James Watson warned in the *Science* article quoted earlier in this chapter, the danger of a move 'toward a more nationalistic approach to science . . . counter to the traditions that have allowed me to admire and enjoy the scientific life', seems already to have become a reality, and he has been its most illustrious victim. Watson's hope, that the human genome should belong to the world's people as opposed to its nations, seems doomed to failure. Yet, in comparison with some of the other consequences which will flow from the Human Genome Project, the issue of patenting is comparatively minor. The way in which this matter has been handled does not inspire confidence that other, less tractable issues will be dealt with properly. The first of these issues will be that of diagnostic tests for genetic disease. This is examined in the next chapter.

6

Mosquitoes and Morals

THROUGHOUT its history the human race has been blighted by the blood-sucking malarial mosquito. The bite of an infected anopheles mosquito – always the female, for males feed only on plant juices and so do not spread disease to humans – injects microscopic parasites which find their way to the victim's liver where they quietly reproduce. After two weeks, the infected liver cells burst open and the parasites spew out to invade the red cells of the blood stream which the parasites then consume from within. When the destroyed cells burst, they release yet more parasites into the blood stream. Infection is signalled by weakness, followed by fever, headache and muscular pains; without early treatment, a cycle of teeth-chattering chills and shakes accompanied by high fever and splitting headaches can set in. There are four types of malaria parasite, and in those victims infected by the most harmful, *Plasmodium falciparum*, the malaria may progress to sudden coma – from which one in five never recovers.

The global death-toll from malaria is unknown, but in Africa alone more than 800,000 children die from the disease every year. The disease is endemic not only in Africa but throughout the tropics and subtropics. Each year there are about 110 million new cases throughout the world. Its baleful influence stretches across the Middle East and Asia to Indochina and Indonesia. Large areas of Central and South America are also malarial. Until comparatively recently, it was widespread in Europe and it pervaded the Mediterranean region until modern times: Cyprus and southern Italy were malarial until after the Second World War, when mosquito eradication programmes removed the carrier insects. The disease needs a fairly dense population to keep it going – it is transmitted when an uninfected mosquito bites a malaria victim, ingests *Plasmodium* along with the blood, and then reinjects the parasites into an uninfected human at her next blood meal. The disease probably began to attack humans about

97

10,000 years ago when Africans switched from hunting to farming – a mode of life which can support a much greater population on the same area of land than can hunter-gathering.

But what has the history of malaria got to do with human genetics? The answer is that in a classic example of evolution in action over many thousands of years the peoples who live in malarial areas have acquired some inborn defence against the parasite, defences which have been transmitted down the generations. This may sound superficially like benign Mother Nature conferring protection on humanity, but in fact the evolutionary response to malaria is a textbook illustration of how natural selection is indifferent to what humans would consider moral and ethical concerns: Nature is literally inhumane. At least three defences have evolved, but they all have side-effects which have been responsible for great human suffering. Collectively they represent a prime example of what the late Sir Peter Medawar called 'a profound truth that Nature does not know best; that genetical evolution is a story of waste, makeshift, compromise and blunder.' The three most common single-gene defects to afflict humanity appear to result from this evolutionary response to malaria. Two of them, sickle-cell disease and thalassaemia, affect haemoglobin. The third also affects the red blood cells, but represents a deficiency in another protein – glucose-6-phosphate dehydrogenase (G6PD).

Collectively, the evolutionary response to malaria illuminates the question of whether there are such things as 'good genes' or 'bad genes'. The subject of this book is the moral and social consequences of 'the new genetics' – a term coined in 1980 by David Comings, the editor of the *American Journal of Human Genetics*, when commenting on the new techniques for analysing DNA. The purpose of exploring the possible moral and social impact of any new technology is to point up some consequences which may be possible but are undesirable and to do so before such consequences are realized in practice. The problem is that the new genetics is new; it will alter our lives but it has not done so yet. At the Hastings Center in New York State, the most prestigious institute for the study of the ethical problems in biology and medicine in the United States, there is a programme of study into the effects of the Human Genome Project. As the Center's director, Daniel Callahan, told me, the work will be 'difficult because it's so hypothetical. You have to use existing precedents.' And for some of the most important consequences of the human genome programme, he believes, 'We don't have precedents or ways of predicting what people or institutions will do with the information.'

If so much of the outcome is hypothetical, readers of this book might well ask whether the social and moral consequences of the new genetics can only be judged after they have happened. If they are discussed in advance, is there not a danger of the author setting up his own Aunt Sallies? How many of the fears are likely to be real? The best way of avoiding such difficulties is, as Callahan suggests, to look at past experience. There was human genetics before the 1980s. Any hypotheses about the social and moral impact of the new genetics of DNA analysis will rest on firmer ground if the lessons to be learned from this older phase of human genetics are first looked at.

The evolutionary responses to malaria occupy prime place both in the history of human genetics and in the story of how modern society has treated those who suffer from inherited disease. These conditions, and sickle-cell anaemia in particular, thus have an interest which far transcends the professional concerns of geneticists: because they were identified early, the way in which society has dealt with them may point to the way in which we deal with the plethora of information that we can expect from the Human Genome Project. In the industrialized countries, moreover, sickle-cell anaemia and G6PD deficiency tend to affect a minority racial group, people descended from black Africans, so to the problems of dealing with an inherited defect are added the complications of racial disadvantage.

The story is not an edifying one. If Nature seems wasteful and blundering in creating these diseases, humanity has not shown itself much wiser in dealing with them. Many of the poorer countries of Africa are unable even to provide enough food for the population, let alone personal medical care: babies born with sickle-cell anaemia simply die. But even in the richest country in the world, sickle-cell disease has been badly handled by society. In the 1970s, many states in the USA instituted screening programmes for sickle-cell anaemia, which did little to help those afflicted with the condition but which resulted in discrimination against healthy people who happened to carry only one copy of the gene. Even though these people were not affected by disease at all, some were denied employment and health insurance on the basis of their genetic make-up.

Second, although sickle-cell anaemia is the world's second

commonest genetic disease, and its cause had been known since the mid-1950s, drug companies have not thought it sufficiently profitable an area to invest time or money in developing drugs to treat sufferers. Although a 'cure' is beyond medical technology at present, it might have been possible to develop treatments to make sufferers' lives as bearable and near-normal as possible. The example of sickle-cell anaemia stands as a serious counter to one of the main claims made by the proponents of the Human Genome Project: that by reading humanity's genes we will speed up the process of developing cures for genetic diseases.

In the case of the thalassaemias, whose underlying causes were teased out somewhat later than for sickle-cell anaemia, screening programmes have been more successful. In some Mediterranean countries, the availability of prenatal diagnosis has greatly decreased the incidence of the disease and in Britain there has been a successful programme of prevention of thalassaemia among the Cypriot population in London. This may raise moral qualms for some, because the only 'treatment' which can be offered in response to a positive prenatal diagnosis is termination of an affected pregnancy. However, the situation is more complex than it seems, and the availability of prenatal diagnosis may actually decrease the abortion rate within communities affected by genetic disease.

Finally, these diseases raise the issue of international equity. Because they predominantly afflict the inhabitants of the poorest countries of the world, and minority ethnic groups in the developed countries, they are important touchstones for how the proceeds of the Human Genome Project are to be shared out. Are the peoples of the Third World going to be remembered in the choice of study by scientists in the First World, or will ailments such as heart disease and late-onset diabetes, which are genetically more complicated and therefore scientifically more interesting, attract the researchers' attention? Such disorders tend to be diseases of affluence which disproportionately affect the rich.

Once a genetic defect has been pinpointed, the first application of the new knowledge is likely to be to devise tests to pick out those individuals who carry the mutation and who are therefore likely to be at risk of suffering from the defect, or of passing it on to their children. Such genetic diagnosis is a very new development with

far-reaching implications. Two considerations are particularly prob-
lematic: first, genetic diagnosis will be available long before there are
any cures for the diseases diagnosed; and second, the diagnosis may
be made long before any symptoms show up. The full force of this
latter consideration will not be felt for some time. At present the only
late-onset diseases for which there are genetic tests are comparatively
rare; however, as researchers increase their understanding of polygenic traits and develop tests to disclose an individual's susceptibility
to such a condition, heart disease for example, then the problems
attending advance diagnosis will come to the fore. There are many
instances at present where doctors diagnose diseases that they cannot
cure: many cases of cancer are sadly incurable but in such circumstances disease has usually set in and the patient is already feeling ill
before the diagnosis is made. The only comparable ailment where
patients who feel perfectly healthy are likely to be told years in
advance that they may die of an incurable disease is AIDS. The
experience with AIDS suggests that genetic diagnosis may require
very careful scrutiny.

The availability of a genetic test for a human disease highlights
three issues. One is screening the population to pick up at an early
stage those who have a genetic defect which may not yet have made
itself manifest, for example screening young adults for a late-onset
disease such as Huntington's. Is there any point in the pre-
symptomatic diagnosis of a condition for which there is no effective
remedy? The second major issue is screening the population to pick
out carriers – those who carry a single copy of a recessive gene and
who therefore do not suffer from the disease but are at risk of passing
it on to their children. Here the question is one of giving individuals
information which will allow them to make a fully informed choice
when they come to consider starting a family. The third issue is that
of treatment: who will develop treatments for such diseases; who will
pay for the treatments to be developed; who will pay for them to be
administered, if any are available? This last is an economic and
commercial issue which is discussed in the next chapter. The rest of
this chapter will discuss the issues of genetic screening, focusing first
on the example of screening for sickle-cell anaemia in the United
States in the 1970s. We shall then look at the studies in Britain in the
early 1990s of prospects for screening the population to identify
carriers of the cystic fibrosis (CF) gene, for the story of these British

pilot studies suggests that at least some of the lessons of sickle-cell anaemia have been learned. Then the imponderables of screening for late-onset diseases will be discussed. This chapter will also address the issue of who should have access to the results of genetic tests.

First, it is worth getting these diseases into perspective. The toll of human suffering exacted by sickle-cell disease and by thalassaemia is enormous. Sickle-cell anaemia predominantly afflicts Africans, with a frequency of about 1 in 50 to 1 in 100 births. In contrast, the most common single-gene defect to afflict northern Europeans, cystic fibrosis, has a frequency of 1 in 2,500 births – some fifty times less. Each year, about 100,000 infants are born in Africa with sickle-cell disease. The disease crops up among families of African descent who have long since left (in the case of slaves, been transported away from) malarial regions: there are about 1,500 cases a year in the USA, 700 in the Caribbean, and about 140 in the UK.

Thalassaemia tends to affect peoples living around the Mediter- ranean Sea and also Orientals, sometimes with a frequency as high as 1 in 50 to 1 in 100 births. According to Sir David Weatherall, 'The global data for the frequency of the thalassaemias are staggering.' In Thailand alone, Sir David records, more than 500,000 children, out of a total population of 50 million, 'suffer variable degrees of chronic ill-health due to the interaction of the different thalassaemia genes.' It seems likely, he notes drily, that the disease is equally widespread in parts of southern China, Laos, Cambodia, Malaysia and in areas of the Indian subcontinent. Thalassaemia also represents a major health problem in Italy, Greece, Cyprus and Sardinia. According to an estimate made in 1979, Italy had about 4,000 severely affected child- ren. The World Health Organisation estimated in 1983 that about 242 million people around the world carried a copy of a gene for thalassaemia or sickle cell.

The third common condition affecting red blood cells, G6PD deficiency, does not give rise to chronic ill-health but men who carry this gene may suffer from severe anaemia in response to some drugs – especially modern antimalarial drugs – or environmental chemicals. G6PD deficiency affects about 100 million individuals worldwide.

Patients with sickle-cell anaemia have a form of haemoglobin that differs very slightly from normal. The result is that once the haemo- globin delivers its oxygen, the molecules tend to stick together and aggregate into protein fibres. This causes the red blood cells to become rigid and sickle-shaped. Such cells do not survive in the circulation as long as normal red cells. They either break down or get trapped in narrow blood vessels. The result is anaemia – through loss of red blood cells – and damage to organs as the small blood vessels and capillaries get choked with the sickle cells and the cells become starved of oxygen. The result of this logjam in the blood stream is tissue death and pain 'crises' – the patient experiences severe pains in the joints, back, chest or abdomen which can last a few hours or up to three weeks. Some patients get crises once a month, others much more rarely. People with sickle-cell disease are also more susceptible to infections. Blood transfusions and bone marrow transplants (if a suitable donor can be found) are the only treatments. Thalassaemia too leads to severe anaemia which can be compensated for by frequent blood transfusions or by bone marrow transplantation. If given regular blood transfusions, thalassaemic children can grow and develop normally and are often healthy for their first ten to fifteen years of life, but they tend to die before they reach the age of twenty from the effect of too much iron derived from the transfused blood. Methods of removing the iron have been developed only recently but even so, the most commonly used iron-removing drugs have to be slowly infused into the body over several hours each day, in a tediously painful procedure.

How could such a toll of disease, suffering and death come about as a response to malaria? Although the exact mechanism is not known, the malarial parasite appears to find it difficult to live and multiply inside red blood cells which contain both normal and sickling types of haemoglobin. So someone with a mixed inheritance is better able to resist the effects of malaria. Children only suffer from sickle-cell disease if they receive a copy of the gene for abnormal haemoglobin from both their mother and their father. If someone carries just one copy of the gene, their red blood cells make both types of haemoglobin, very few cells sickle and the carrier does not suffer from the disease. When such a mutation first appeared in human evolutionary history, nearly every sickling gene would be partnered by a copy of the unchanged gene, conferring protection against malaria and leaving the carriers in good health. But as the gene became more common, it became more likely that two carriers would marry and have children. The usual genetic lottery then comes into play. There is one chance in four that the child of such a marriage will be completely free of the sickling gene; if this child lives in a malarial area, every mosquito bite is potentially a sentence of death. There is a 50:50 chance that the child will have mixed genes like its parents, in which case it will have some protection against malaria. And there is a one-in-four chance that the child will receive a copy of the sickling gene from both parents and so suffer the crippling anaemia, which, untreated, leads to an early and painful death. The process of evolution has arrived at a cruel calculus of death such that the numbers of children who die from sickle-cell anaemia and of those who die unprotected from malaria are balanced by the enhanced survival of those with both forms of haemoglobin.

Unravelling the cause of sickle-cell anaemia was the first triumph of what might now be called the 'old' human molecular genetics. The clinical symptoms had been recognized by a Chicago doctor, James Herrick, who in 1910 first observed the characteristic sickle-shaped cells while examining a blood smear from a black West Indian medical student. But it was not until 1949 that researchers identified the underlying cause as an alteration in the molecular structure of haemoglobin. In 1949, Linus Pauling, Harvey Itano and their colleagues showed that sickle-cell haemoglobin moved faster under the influence of an electric field than did normal haemoglobin when placed in a type of gel used for separating proteins. This different behaviour under gel electrophoresis showed that the defect lay within the structure of the molecule itself. The scene then moved to Cambridge, England, where Vernon Ingram was analysing the amino acid sequence of haemoglobin, applying methods worked out a few years earlier by Fred Sanger on insulin. In 1956, Ingram showed that sickle-cell haemoglobin contained just a single change in one amino acid in the sixth position of the β chain of the molecule, where glutamic acid was replaced by valine. And from this single change all the consequences of sickle-cell anaemia arise.

In the case of thalassaemia the defect is generally not in the structure of the haemoglobin molecule but in the amounts of α or β chains made. In some cases no α or no β chain is made. There are over 60 different mutations, giving rise to a variety of thalassaemias. One particularly severe form, known as β-thalassaemia, results from a deletion of the β gene – part or all of the gene has been 'lost' from the person's DNA, and no β-chain is made. More often, however, the thalassaemias are the result of a single base change. As in the case of sickle-cell disease, those with mixed inheritance are protected against the effects of malaria. The evidence is clear: for example, in the Pacific islands, α-thalassaemia is very frequent in Papua-New Guinea whereas it is uncommon in the south, in New Caledonia, and the distribution of malaria exactly mirrors that of α-thalassaemia in these islands.

Sickle-cell anaemia was rediscovered as a major health problem in the USA in the late 1960s and at the beginning of the 1970s. The black population was told that many of them were suffering from this hitherto neglected disease. The incidence is about 1 in every 400 blacks and about 1,500 black babies are born with the disease each year in the USA. Concern over sickle-cell disease coincided with the growth of the civil rights movement in the USA and followed the introduction, in the early 1960s, of mass screening of all newborn babies for a different genetic disease known as phenylketonuria (PKU). Sickle-cell is a more common condition than PKU which affects only 1 in 12,000. In many cases at the behest of black lobby groups and black legislators, individual states within the USA started programmes to screen for sickle-cell disease, modelled on PKU screening. Eventually, with financial support from the Federal Government, a national screening programme was set up. This was long before the advent of techniques for DNA analysis so the screening tests were based on analysis of haemoglobin in blood samples taken from infants and children. But the programme was a fiasco and quickly became the focus of political and racial contention.

The first problem was that the two diseases are not comparable. If nothing is done about PKU, it leads to mental retardation and early death, but the defect is comparatively simple to deal with. Those who have the defect cannot digest an amino acid normally present in their diet, but the effects of the defective gene never become manifest if those with PKU receive a modified diet in which none of this amino acid, phenylalanine, is present. With sickle-cell anaemia, there was no way of dealing with the disease: diagnosis would not lead to treatment or cure. Parents of children with PKU needed and were given advice and counselling on how to cope with the defect and on how to restrict their child's diet. The sickle-cell anaemia screening programme was not backed up with genetic counselling, so very few of those couples who each carried the gene and were at risk of having children who suffered from the disease received proper advice.

Also, there was at the time no way in which an unborn foetus could be tested in its mother's womb with a view to offering the mother the option of not continuing with an affected pregnancy. (Abortion was in any case illegal in the USA until 1973.) None the less, several states made screening of schoolchildren or of newborn infants compulsory, without providing education or counselling for the affected communities.

Worse still, some people were tested to see if they were carriers without a clear distinction being made between those carrying the trait (having only one copy of the gene) and those actually suffering from the disease. As a result, many parents of children with sickle-cell trait falsely believed that their children had a debilitating disease or at least that they ought to restrict what their children did so as to prevent the illness taking hold. The situation was not helped by remarks made by Linus Pauling in 1968. These were capable of being interpreted as suggesting that carriers of the trait should be 'branded' so that they would not marry each other or at least not have children. Confusion between sickle-cell trait and sickle-cell disease was widespread and was reinforced by the media. At one point in 1972, an advertisement placed in the magazine *Ebony* (which has a predominantly black readership) and which was intended to raise money for research into sickle-cell disease, falsely characterized carriers of the trait as being weak:

It's a killer. One out of every ten Black Americans carries a blood trait that threatens to cripple or kill. It's called sickle cell disease, because it creates deformed 'sickle-shaped' red blood cells. It can weaken those it doesn't kill. Even those with the milder form of sickle cell disease – the 'trait' – suffer. Usually they must avoid strenuous activities and consult their doctors on a regular basis.

The advertisement was sponsored by American Express, who offered to donate the purchase price of American Express Money Orders to the fight against sickle-cell disease.

The confusions abounded. In Washington DC, sickle-cell anaemia was made in law a communicable disease, and schoolchildren were subject to compulsory testing under the city's immunization regulations. In the state of Massachusetts, sickle-cell trait became a disease, as far as the law was concerned. According to Loretta Kopelman, writing in the Autumn 1978 issue of *Perspectives in Biology and Medicine*, the test used in the state of New York detected carriers of the gene as well as those who suffered from the disease and, although the law required only that the occurrence of the disease be recorded, if someone carried the trait this was recorded against them also:

We do know that this information becomes part of the permanent medical hospital record of the infant. In the past, insurance companies have demanded as a precondition to insuring individuals, full access to medical records and, upon discovering conditions such as sickle-cell disease or sickle-cell trait, have denied insurance to individuals.

Further discrimination resulted in the job market. Blacks with the trait but not the disease were denied jobs on airlines and even entry to the US Air Force Academy because it was wrongly believed that their blood would react badly to reduced air pressures (and therefore oxygen concentrations) experienced when flying in high-altitude passenger or military aircraft. Only in 1981 did the Air Force Academy lift its restrictions.

Nor were the problems confined to people's professional and working life. Genetics is not about individuals, it is about families; the genes that any individual carries reflect their family history. The screening programme for sickle-cell disease reached into private and personal lives, even threatening marriages. In 1972, Robert Murray recounted to a Hastings Center Conference some of the issues that testing for sickle-cell anaemia had thrown up. Two of his case studies are of particular interest. One was a black mother whose one-year-old child suffered from anaemia and failure to thrive. The child was tested and found to have sickle-cell anaemia. Both parents were tested, but only the mother showed up as having the trait. On being told that the tests were inconsistent with her common-law husband being the father of her child, the woman swore 'on her mother's grave' that he had to be the father. Fortunately, the doctors believed her, and on carrying out a second test they discovered that her husband's red cells did contain a very small amount (about 5 per cent) of sickle-cell haemoglobin, so that he was indeed a carrier.

The case illustrates one fundamental problem of all genetic tests:

there is a non-negligible chance that the test will actually be wrong. In this case, it was a false negative: if the doctors had believed their technology rather than their patient, it could have led to the disruption of the family as a result of the father wrongly believing that the child was not his own. Dr Murray's judgement as a doctor was that 'where there is a conflict between human relations, well-being and total scientific truth, I have tended to sacrifice the truth.' He recounts the case of another couple with a sickle-cell child where it became clear that only the mother was a carrier of the trait (and therefore that some man other than the non-carrier who believed himself to be the father must have fathered the child). According to Murray, 'It was explained to them that an egg from the mother containing a sickle-cell gene was fertilized by a sperm in which a fresh mutation also producing a sickle-cell gene had occurred. It was not pointed out that mutations are extremely rare.'

At this time several states, including Kentucky, Rhode Island, South Carolina and New York, introduced laws requiring that newborn infants be tested for sickle-cell disease. But in the 1970s, treatment for sickle-cell disease consisted mainly of relieving the pain crises. Testing of newborns was pointless as palliative treatment could in any case be given when the first symptoms presented. Indeed knowledge that a child had the disease may have led to medical misdiagnosis. A report by the American medical ethics research institute, the Hastings Center, cites one documented case where a child died because the doctor – who had been told by the mother that the child had sickle-cell disease – thought he was seeing a sickle-cell crisis. In fact, the child was a carrier, did not have the disease, and died of acute appendicitis. Given the problems of stigmatizing infants and burdening them with medical records that would blight future employment prospects, the disadvantages of screening newborns outweighed any advantages. In 1978 Professor Kopelman concluded her review by warning that proper account should be taken of the risks involved in screening for genetic diseases and that risk must be thought of as something more than the physical risk of injury at the time of testing. 'It involves the risks of labelling, the invasion of privacy, loss of self-esteem and discrimination. No one has the right to subject others to procedures which on balance involve risk to them.'

It was not even as if widespread testing was required for

epidemiological purposes – to assess how widespread the disease was
in the community. That end could be more easily achieved by small-
scale sampling, provided it was statistically well controlled. Any testing procedure aimed simply at compiling figures for the incidence of a
disease has to reach a balance between the interests of the individuals
being tested, or likely to be in the same group as those tested, and
wider society's interest in the knowledge so gained. Should there be a
conflict, most democratic societies tend to place greater emphasis on
the interests of the individual unless the collective need for knowledge can be shown to be of paramount importance. The premium to
be placed on the individual's rights seems to be all the greater when
the issue is something as personal and intimate as their genetic
constitution and their right to bear children. There are too many
historical precedents of great human suffering resulting when individual rights have been over-ridden and the state takes it upon itself to
dictate who will and who will not have the right to have children.

   The sickle-cell testing programme in the USA eventually col-
lapsed in disarray: at one point seventeen states had mandatory
testing on their statutes whereas a decade later, only ten continued to
offer screening. The unhappy saga was one of the stimuli for a 1983
report on genetic screening, drawn up by a special advisory body to
the US President. The President's Commission for the Study of
Ethical Problems in Medicine and Biomedical and Behavioural
Research, to give it its full and somewhat unwieldy title, made several
important recommendations in its report 'Screening and counselling
for genetic conditions: A report on the ethical, social and legal
implications of genetic screening'. The first recommendation was for
confidentiality: unrelated third parties should not be given genetic
information. The second was to advise against mandatory screening
programmes: their scope was limited, the commission said, and
voluntary programmes would always be preferable except where they
proved inadequate to prevent harm to defenceless individuals, such as
newborn babies. The commission also recommended that doctors
should, so far as possible, tell their patients the truth about the results
of any test: honesty between doctor and patient was indispensable.

   Moving from the position of the individual to the social and public
policy implications of genetic testing, the commission made the point
that mass-screening programmes should only be set up if well-
conducted pilot studies had already been undertaken and had

demonstrated that the screening would actually be worthwhile. Counselling and follow-up care should be integral parts of any screening programme, and screening should not proceed until such arrangements had been made. Access to screening ought to take into account the frequency of different genetic diseases within different subgroups or racial groups of the population.

In April 1987, a panel of experts set up by the NIH once more recommended that newborn screening for sickle-cell disease be offered to all babies. This time, the motivation was different. Clinical research had shown that children under the age of three with sickle-cell anaemia are less able to fight bacterial infections and have a 15 per cent chance of dying from infection in the first few years of life. The research showed that this could be prevented by administering penicillin to the children as a prophylactic. The NIH recommended that once children were diagnosed as having sickle-cell disease, they should receive penicillin from the age of four months until they were five years old. Screening now had a purpose, and could deliver some benefit to those individuals being screened: protection against previously fatal infections.

Two points are worth noting about this research. The first is timing: the underlying causes of sickle-cell anaemia had been known since the mid-1950s, screening programmes had been introduced in the early 1970s, yet it was only in the late 1980s that conclusive research had been done on so obvious a topic as how well babies with the condition thrive. The second point is the irrelevance of genetics: the research consisted of observing the welfare of sickle-cell babies and comparing them with babies without the condition. Increased susceptibility to infection in the early years of life was observed; it was not deduced from knowledge of the genes. Although the Human Genome Project will provide new knowledge about human genes and human disease, its results will not be definitive: as the case of sickle-cell disease illustrates, reducing human beings to their most elementary parts, the genes, does not tell you everything about human beings. There will still be extensive scope for clinical and other nongenetic research to reveal important things about us which are not written in our genes.

Some of the states which still require compulsory screening seem to have learned the lessons of earlier errors. In California, where newborn screening for sickle-cell disease is still compulsory, parents

counseling

are given extensive counselling and asked to attend family education sessions every two to four weeks in the first few months after the birth. Parents are taught to recognize danger signs and are given a letter to carry with them when they travel to inform doctors that the child has the disease and to provide basic medical information. Such letters are necessary, for many doctors who practise in areas with few African-Americans may never have seen a case of sickle-cell disease and may misdiagnose the symptoms. To counterbalance the case of appendicitis misdiagnosed as sickle-cell disease mentioned above, there are documented cases of children in the midst of a sickle-cell crisis being admitted to hospital where doctors unfamiliar with the disease promptly accused the parents of deliberately injuring their child.

Developments in medical technology have changed the options available to those at risk of passing on genetic disease, compared with the situation at the start of the sickle-cell anaemia programme. Since the early 1970s, it has been possible for a doctor to tell a mother if the child she is carrying is affected by some genetic and other defects. Prenatal diagnosis is founded on ways of securing samples of cells from an unborn foetus – even the foetus's blood can be examined.

Amniocentesis was the first procedure to be developed. Cells floating in the amniotic fluid surrounding a foetus belong to the foetus and not the mother. They can be extracted and analysed to detect chromosomal abnormalities and some biochemical defects. The doctor sucks out a small amount of the amniotic fluid through a needle inserted through the mother's abdominal wall; the doctor guides the needle by looking at ultrasound images of the location of the placenta and the foetus so that it does them no injury when it enters the amniotic sac. But before DNA-based diagnosis, the range of disorders which could be detected by amniocentesis was fairly limited. Neither sickle-cell anaemia nor thalassaemia, for example, could be picked up, because the cells in the amniotic fluid do not make haemoglobin. In 1974, however, researchers found a way of sampling foetal blood, which did enable screening for defective blood proteins.

The disadvantage of both these techniques is that they can only be done comparatively late, after the fourteenth week of pregnancy. Moreover, amniotic cells must be grown on in the laboratory before there is enough material for analysis and this may take a further three weeks. Such a long time lag makes it very hard for the parents to decide to terminate the pregnancy if the foetus is affected with some severely disabling disorder, and such late terminations are physically and emotionally very distressing for the mother.

A different technique known as chorionic villus sampling (CVS), which can be done earlier in pregnancy, has been developed more recently. Chorionic villi are fingerlike projections of the membrane surrounding the embryo in early pregnancy which gradually evolve into the mature placenta. CVS can be carried out during the eighth to the eleventh week of pregnancy.

Like amniocentesis, this technique yields cells that can be examined under the microscope for gross chromosomal abnormalities or whose protein content can be analysed to check for some inherited disorders. Protein analysis can pick out a case of Tay-Sachs disease, for example. This fatal disorder of the nervous system results from a deficiency of the enzyme hexosaminidase A. It can be detected because normal amniotic cells contain hexosaminidase A, whereas cells from affected foetuses do not. But none of these techniques could pick out cases of PKU deficiency, for example, because the enzyme is produced almost exclusively by liver cells which are not collected by amniocentesis, CVS or foetal blood sampling.

The advent of direct DNA analysis has changed all that, enabling a search through the chromosomes themselves for the DNA patterns characteristic of the disease state. In 1978, Y. W. Kan and Andrees Dozy at the University of California, San Francisco, reported that they had conducted an antenatal diagnosis of sickle-cell anaemia by analysing the DNA in the cells of the amniotic fluid. This was the first prenatal diagnosis using DNA analysis. Couples who knew they were at risk of transmitting an inherited disease – perhaps because they already had an affected child – could now be offered a prenatal test early in a subsequent pregnancy to be assured that the foetus was free of the disease and, if it was not, to be offered the option of termination. In principle, the options extend even further, so that those couples with no previous history of the disease in their families, but who could be identified as carriers by mass population screening, could also opt for prenatal diagnosis and, if they wished, elect for termination of affected pregnancies without ever having given birth to an affected child. This option is now moving out of the realm of

theory and into practice for the detection of cystic fibrosis, as we shall
see later.

The sickle-cell episode was a deeply unhappy one, but there have
been success stories in the USA, most notably with Tay-Sachs
disease. This occurs comparatively commonly in Jewish families of
European ancestry, with the incidence varying from 1 in 6,000 children to as high as 1 in 2,500. (The incidence is something like 1 in
600,000 for non-Jews.) Like PKU, it is an inborn disorder of meta-
bolism: those who carry two copies of the defective gene lack the
enzyme hexosaminidase A, which results in the build-up of fatty
deposits on the nerve cells. The symptoms show themselves within
the first four or five months of life and the condition leads to mental
retardation, blindness, paralysis and dementia, with most affected
children dying by the age of four. The number of infants born with
Tay-Sachs disease in the USA has dropped from around a hundred a
year in the early 1970s to fewer than ten a year in the late 1980s,
owing to a properly co-ordinated programme of prenatal screening
accompanied by the option of aborting an affected foetus.

In the United Kingdom, Dr Bernadette Modell and her colleagues
from University College Hospital Medical School have established a
notably successful programme of prenatal screening for those couples
at risk of having a child with thalassaemia. There is a large Cypriot
community in London whose members are affected by this disease.
Before prenatal diagnosis became available, couples who knew they
were at risk had virtually stopped having children. These couples had
either already had a child with thalassaemia or had both tested
positive in carrier screening tests. Most subsequent pregnancies
among such couples were accidents and 70 per cent were aborted
because the prospective parents felt that they could not run the
one-in-four risk of having an affected child. The introduction of
prenatal diagnosis changed the situation dramatically. The abortion
rate went down: as one might expect from the genetic statistics, fewer
than 30 per cent of subsequent pregnancies needed to be terminated
because of a prenatal diagnosis of thalassaemia.

However, there have been failures. Immigrant Asian communities
in Britain have been much less willing to take up genetic screening as
a means of decreasing the incidence of thalassaemia than the Cypriot
community. Another project, offering genetic screening for thalassaemia to Greek villagers, ended in the distress of those the project

was intended to help, with some villagers having been stigmatized. This happened even though the project was well designed and adequate genetic counselling was offered. Moreover, there was no decrease in the prevalence of the disease.

Even where a particular community disapproves of aborting an affected pregnancy, there can sometimes be other ways of decreasing the incidence of genetic disease. Among the Hasidic Jewish population of Brooklyn, New York, for example, couples tend to enter arranged marriages. In 1984, a programme began, with the co-operation and support of local Hasidic organizations, to test the population for carriers of Tay-Sachs. The results of the tests were made available (with due safeguards for confidentiality) to the community's marriage brokers, so that if two potential partners were discovered to have the gene, the marriage was not proceeded with. As the data were kept confidential, there was no possibility of public stigma of the individuals concerned. The result has been a considerable drop in the number of births of infants with Tay-Sachs without recourse to abortion. However, this example is surely exceptional.

Confidentiality

When scientists in Michigan and Toronto discovered the main mutation leading to cystic fibrosis (CF) in 1989 (see Chapter 2), this opened the way to a test for the presence of the gene. In principle, the entire population could now be screened to pick out those carrying the mutation. Although such mass screening may appear analogous to that for sickle-cell anaemia in the 1970s, the focus of CF screening is different. While the American sickle-cell screening programme sought to identify those suffering from the disease and picked up carriers of the trait only as a side-effect, the point of CF screening will be to identify carriers – more specifically, couples who unknown to themselves are at risk of having children with the disease. Children who suffer from CF itself were of course being diagnosed by other procedures long before genetic analysis became available.

Although American researchers played a significant role in discovering the defective CF gene, with the country's mixed legacy from past genetic testing it is not surprising perhaps that the United States has been slow to apply the knowledge generated by the new genetics. In addition, abortion remains a highly charged emotional and political issue in the United States. A third factor is that the system for delivering health care in the United States is peculiarly unsuited to

U.S.: Response to finding CF Gene

providing cheap and effective mass screening, so that any CF carrier screening programme is likely to be expensive. Finally, because health care is funded by private insurance companies, those who test positive are likely to find themselves discriminated against or at least having to pay higher premiums to secure health cover and this may prove a powerful economic disincentive, making most individuals reluctant to have themselves tested. Thus it was not until April 1991 that the NIH made money available – about $1 million – for pilot studies to evaluate how best to deliver a DNA test for carriers of CF.

In contrast, pilot studies to screen for carriers of the CF mutations were being discussed in Britain within two months of the announcement that the gene had been isolated. By December 1990, trials had been set up by the Medical Research Council (MRC) together with the British Cystic Fibrosis Research Trust. All newborn infants in Britain are screened for PKU, but the Council and the Research Trust considered and rejected the idea of screening for CF carriers at that time. Instead they decided to target people near reproductive age or in the process of having a baby. Writing in the MRC's newsletter, D. J. H. Brock, professor of human genetics at Edinburgh University, cautioned that 'Any type of screening has the capacity to disturb as well as to reassure. Being told that they are a carrier of a CF gene may be quite upsetting to some people however carefully they have been prepared for this possibility.' Consequently, the British carried out six trials to discover the best model for delivering mass screening.

The idea of screening for carriers is to give people fuller information about their genetic status and allow them to make a fully informed choice when they make decisions about reproduction. There are a number of options for those who know that they carry the CF mutation. They may use the knowledge to guide their choice of partner and marry only someone whom they know is not a carrier; or they may marry regardless and, if they discover that their partner is also a carrier they may then decide whether or not to have children. If they do want children, they could choose to have them by using sperm or eggs donated by someone certified not to be a carrier, or they could adopt a child. A further choice is to start a baby in the normal way but then have a prenatal test to see if the foetus is affected and, if so, opt for early termination. A final informed choice is of course not to act on the information at all, but to have children without making use of the results of any genetic tests.

Options of CF Carriers

The availability of so many options is something very new in human experience – perhaps only couples who wish for but cannot have children have faced anything similar in terms of making decisions about reproduction. To some observers, it might appear that there is a general social benefit to be derived from these new choices: the technology may force us all to be more mature and responsible in considering the genetic legacy which we leave behind us. On the other hand, one cannot help but think that technology is outstripping the way people feel about each other and that few people order their personal and sexual lives by such logical choices. There is no place in such a calculus of risk for the feckless, the concupiscent, or those who in the heat of the moment are merely careless. Above all, this is a passionless computation, whereas most of us might like to believe that passion has a proper place in matters of sex and reproduction.

Professor Bob Williamson, who leads one of the groups conducting a pilot study, says, 'Choice carries with it real responsibilities and not everyone is happy with them. Information can be liberating, but it can also be a burden.' If you test positive as a carrier, 'what do you do?' Williamson continued. 'Do you even tell your partner? What about someone who does not have a partner? Do you tell them when you first meet, or when you first go to bed together? All of a sudden you are throwing on to people with no previous experience of the condition an enormous amount of information.'

Most of the British trials took advantage of one of the characteristics of the British medical system: that everyone is registered with a local family doctor or general practitioner (GP) to whom they go first if they feel ill or need treatment. Supplementing the local GP, in many areas there are family planning clinics to which people can go for contraceptive and other advice. Almost all women who become pregnant attend antenatal clinics and receive continuing antenatal care either from their GP or as out-patients at local hospitals. All these services are available free of charge under the country's National Health Service, so there is no financial disincentive to taking up any of them.

Three of the CF screening pilot studies made use of the GP and family planning system to offer CF tests either to patients who had gone to see their GP for other reasons or to those who had gone to family planning clinics for contraceptives. In another study in Edinburgh, Scotland, screening was offered in the antenatal clinics of a large maternity hospital. It was recognized that this may not be an ideal moment at which to screen, because in the event of a positive test on the prospective mother, couples have only limited time in which to take the very serious decision to seek a prenatal test on the foetus with the possibility of electing to terminate an affected pregnancy. A further disadvantage is that some women only attend antenatal clinics when their pregnancy is well established, and given the further time gap for the tests to be done this could mean a very late and distressing abortion.

One study at a London hospital targeted only couples, treating them as positives only when both prospective father and mother were carriers of the gene. A further study offered testing through secondary schools. One possible disadvantage of this is that for an adolescent to be diagnosed as a carrier may be an unwanted psychological burden at a time of great emotional turmoil. The studies are still being evaluated to see what the strengths and demerits of each route are. However, some preliminary results have been published. The initial results of the study in which Williamson is participating, of more than 1,000 patients aged from sixteen to forty-four, and carried out in a relatively prosperous area just north of London, found a very high rate of acceptance. The offer of a test was taken up by 66 per cent of those in GPs' surgeries and by 87 per cent of those attending family planning clinics. A third group of people received 'cold-calling' letters offering a test but only 10 per cent accepted the invitation.

The test is simplicity itself and is described in Chapter 4. One problem for CF screening is that the commonest mutation affects only about 78 per cent of those in the British population who have a CF mutation. (The incidence of this mutation varies with different nationalities and different ethnic groups; in Denmark, for example, it is 90 per cent.) Two other mutations can be picked up in the course of the test, which brings the detection rate to more than 82 per cent. According to Williamson, there are no 'false positives': no one who is free of the mutation is wrongly diagnosed as carrying it, because all positives are tested a second time. But there are false negatives. 'You can't identify all the carriers,' Williamson says. 'But that's common in medicine, for example screening for cervical or breast cancer does not pick up all the cases but that does not mean that it isn't worth doing. We can identify more than two-thirds of those couples who are at risk and it is right to offer this information to the two-thirds we can

CF Screening

identify even though there are some we cannot.'

In sharp contrast to the American experience with sickle-cell anaemia screening, the state has not been involved in these British pilot studies. Even though the MRC receives its funds from government, it is constituted in such a way as to keep its research activities at arm's length from direct political control. None the less, there is an economic dimension to the work which is bound to be of interest to the state as ultimate funder of medical care in the UK. Williamson reckons that it costs about £2 to conduct a test, including the price of the laboratory reagents and the technician's time. On top of that there are capital costs of setting up the analytical laboratory in the first place, the costs of collecting the samples and of providing genetic counselling for those who are diagnosed as carriers. The total cost of a screening service is therefore more likely to be several tens of pounds sterling per sample and, given the frequency of carriers in the population, at least £50,000 will have been spent before one case of CF itself is identified in this way.

The question of whether there is any economic benefit to the National Health Service in mass screening turns on the behaviour of at-risk couples. If both partners are carriers, will they elect to terminate a pregnancy where the foetus has been diagnosed prenatally as a sufferer? The cost to the National Health Service of treating someone who has CF is around £10,000 a year. Life expectancy is at least twenty years, and with advances in medical care may be more like thirty to forty, so the average lifetime cost of caring for a CF sufferer will certainly exceed £200,000. The health service will therefore save money only if more than a quarter of those couples in such a situation elect to terminate their pregnancy. If, however, more than three-quarters decide to continue, then the carrier screening programme will have cost more than not screening at all.

Cost related to CF

This may sound like a callous calculation, but the financial resources even of comparatively wealthy nations such as Britain are not endless and money spent on screening for CF is money that cannot be spent on providing medical care for other equally needy causes, so there is an obligation that the money should be well spent. Williamson's initial survey identified twenty-nine carriers of the gene but did not pick up any carrier couples. None the less, the study did ask participants to consider what they would do if both they and their partner were found to be CF carriers: 17 per cent said that they would not contemplate terminating an affected pregnancy, 26 per cent said they would and some 57 per cent were unsure of how they would react. How far these responses would reflect actual behaviour in the event, however, is not clear.

These preliminary results suggest that in purely economic terms screening might be advantageous. One US estimate of the costs and benefits of carrier screening came to exactly the opposite conclusion. The US estimate was published in the *Journal of the American Medical Association* in May 1990, a year before money was available for pilot studies in the USA and nearly 18 months before publication of the British study. The authors, Benjamin Wilfond and Norman Fost, from the University of Wisconsin School of Medicine, estimated the lifetime medical costs for someone with CF as around $200,000 (only about £110,000, half the British estimate) whereas the cost of each test was put at more than $200 (at least five times the British estimate). The American calculation put the cost of avoiding one CF birth at more than $1 million. Under this analysis, screening for CF carriers would impose significant extra costs on the health care system.

The American estimate for the cost of a single test is extraordinarily high compared to the British one. To a large degree this may reflect the different health care systems in the two countries. If the UK did decide to proceed with mass screening, the mouthwash test could be administered by local GPs, to whom could also naturally fall the role of preliminary genetic counsellor. Administration of the test and initial genetic counselling would thus incur essentially no extra cost to the health service. Analysis would be carried out by a DNA laboratory serving the local area and also funded by the National Health Service. In the US, however, fewer people have a primary care physician with a similar role to the British GP, so any test is likely to be done by specialists, which involves more costs. The analysis would be done by private, profit-making companies, and the genetic counselling would also be done by specialists. It is thus likely to cost more to deliver the same service under the American system than the British.

The attitudes of American doctors towards genetics are likely to reinforce this trend. In 1974, a survey of paediatricians, obstetricians and family physicians commissioned by the US National Research Council found that fewer than 2 per cent of respondents believed

Genetic
Counseling

physicians were competent to provide genetic counselling. This is perhaps not surprising given that as recently as 1985 20 per cent of US medical schools did not offer courses in genetics. But only a few of those responding to the 1974 survey thought that physicians should receive more training in genetics: most thought that specialized counselling clinics should be available or that counselling should be done by trained counsellors.

It is worth asking also whether the general population in the USA would take up the opportunity of being screened for genetic disease as readily as the British sample groups say they would. The American situation is complicated by its reliance on private health insurance as the main system of delivering primary health care. Insurance companies exist to make a profit and they do so by selecting those whom they insure on the basis of a statistical expectation that those selected are healthier than the average. If someone represents a higher risk, the insurance companies' response is either to decline insurance altogether or to charge higher premiums.

U.S.
Insur.

Ins.
Discrim.

Screening for sickle-cell disease gave rise to cases of discrimination in health insurance and couples at risk of having a CF child could find difficulties in obtaining insurance in the USA, because the insurer feared the expensive possibility of the couple having a child affected by CF. There is thus a powerful economic motive against having the tests at all. Such rationale against testing may sound compelling. Yet it is difficult to suppress the feeling that a situation in which technology is available but not used to avoid births of children with CF is a perverse one. Provided the tests can be administered in a cost-effective fashion, it ought to be in everyone's interests for couples who are carriers of the CF mutation to be identified. It is in the interests of those so identified, for they then have extra knowledge of their reproductive status and can act, if they wish, to avoid giving birth to a child affected by the disease. Such action would diminish the sum total of human suffering in the future – a morally desirable end in itself. If the testing was cost-effective it would have the practical benefit of decreasing the burden of genetic disease on the overall health care budget, thus freeing money that could be used to alleviate other suffering.

The market orientation of the American economic system is supposed to allow everyone to pursue their own goals, with the presumption that the common good will be best served if individuals are left

free to make their own economic decisions. In fact, in the area of
health care, the issue of genetic disease demonstrates very clearly that
there are situations where, if everyone serves their own needs, everyone is worse off than if they had united to pursue a common goal.
One of the most revolutionary aspects of the Human Genome Project
may be to undermine the system of health care in the USA. It may
turn out that only a socialized health care system, like the British
National Health Service, is compatible with a medical technology
which includes widespread genetic screening.

A primary issue in any plan for genetic testing is the one of privacy
– of who has access to the results of genetic tests. At first sight, it
might appear that some of the problems noted earlier might be solved
by legislation which simply prohibited insurance companies from
gaining access to the results of tests. But as will be discussed later the
issue is not clear cut, and the insurance industry could rightly claim
that such legislation would put them at an unfair disadvantage. Insurers in America have already said that they would wish to have access
to the results of genetic tests in future and, while European companies have not addressed the issue with the same urgency, there is
little doubt that they will reach the same conclusion. If no resolution
can be found, then it is likely that the uptake of genetic tests in the
USA will continue to be low, avoidable human suffering will persist,
and health care will cost more than it might otherwise need to.

Even without such commercial considerations, genetic screening
and selective abortion in themselves create new problems, for if a
given condition becomes increasingly rare there may be less public
interest in finding treatments for it. Much medical research depends
on charity for its funding and charities could find it more difficult to
raise money if there are fewer cases of the disease present in society.
Furthermore, children who were diagnosed *in utero* as being affected
with a genetic disorder but whose parents elected not to terminate the
pregnancy may come to question their parents' decision – the more so
if incidence of the disease is growing rare and those born with it
experience a proportionately greater stigma. 'Wrongful life' suits have
already been successfully brought in the American courts where
infants with birth defects have been awarded damages. In one Californian case of a child born with Tay-Sachs disease, the parents were
awarded damages because the court held that they had negligently not
been told that they were carriers of the Tay-Sachs mutation. As a

result, they had been deprived of an option to abort. The very idea of a wrongful life appears paradoxical, for how can a child be wronged by existing? Depriving someone of life is usually considered to be the wrongful act. The US legal precedents appear really to be a way of gaining compensation for the parents of an afflicted child to allow them to care for it. Wrongful life cases have been rejected by the British courts.

So far, this discussion has concentrated on conditions which are immediate in their onset. But there are other conditions whose effects appear much later. They fall into two classes: single-gene defects such as Huntington's chorea and some rare inherited forms of Alzheimer's disease; and polygenic diseases such as heart disease, diabetes and cancer. Both Huntington's and the inherited form of Alzheimer's are deterministic: anyone who has the mutated gene will inevitably develop the disease, and both are rare. Cancer, diabetes and heart disease, however, are common, and they result both from a subtle interplay between genes and the environment and from the concerted action of several genes together. Except in some very rare inherited cases, there is no 'gene for cancer' and there is no simple cause and effect relationship between the possession of a particular mutated gene and the later development of cancer. Although both classes of disease are late-onset, they present different moral problems.

The first problem is to decide if there is any purpose in carrying out a genetic test for a disease such as Huntington's or inherited Alzheimer's where there is no hope of treatment or even of amelioration of the disease. An adult whose mother or father died of Huntington's will already know that they are at risk of the disease (the chances are 50:50). A test will show almost definitively (with around 95 to 98 per cent probability) one way or the other. That is fine for those who escape the mutation, but it amounts to a sentence of death for those who carry the fatal gene. The knowledge will inevitably cloud their remaining years of health before the disease sets in. Nancy Wexler judges that there are two main categories of at-risk individuals: 'People who are having a good time and have nothing pressing for why they should know, don't come forward for testing. But then there are people who can't stand the ambiguity, who come from the position of theirs being a troubled life for forty years and say "I don't have anything to lose.'' For those who come to be tested as adults, she believes, 'A tremendous lot of counselling needs to go on to see if people need to have the test – many people need counselling more than they need the test. My worry is that too few people are available: there are only about 1,500 counsellors in the US.' In the end, she says, only about one tenth of those who come forward actually have the test. This situation contrasts sharply with the uptake of the CF carrier tests in the UK. But while Americans may well have fears about the effects of a positive test for Huntington's on their chances of getting medical insurance, a low take-up has been a feature of genetic tests for late-onset diseases in the UK too, so personal rather than economic motivation appears to be important.

The family connection provides a second rationale for testing in cases of diseases such as Huntington's. Prenatal testing can ascertain the genetic status of a foetus, to ensure that the gene is not passed on to future generations. If the test disclosed that the foetus had a mutated gene it would indicate not only that the child would suffer from the disease but that the at-risk parent also had the gene, so a positive foetal diagnosis would also reveal the parent's condition. However, if the foetus were free of the mutation, the status of the at-risk parent would still be unknown.

There is a great difference between such cases and prenatal testing for CF and the other immediate-onset diseases. Throughout this book, there has been an unspoken assumption that abortion to prevent the birth of a child handicapped by a genetic defect is morally acceptable, even though some of the world's major religions – the Roman Catholic Church and much of Islam – would not find it so, and even though, in the case of CF, say, many affected children can lead happy and fulfilled lives for many decades with the benefit of modern medical care. However, even those for whom 'therapeutic' abortion is morally acceptable might well pause at the thought of aborting a foetus because it carries the Huntington's gene. If born, such a child could expect a life free of handicap for thirty-five to forty-five years before the disease sets in. Indeed, such are the chances of life that someone with the mutation could die 'prematurely', in a car accident perhaps, and never suffer the disease at all. Is it morally justifiable to deprive a foetus of the prospect of decades of health for fear of the disease later? One answer to this question is that the terms in which it is posed are false. Wexler points out that those

Abortion Right?

who have the gene will have already watched one of their parents progressively deteriorate for more than ten years. They worry that, if they themselves have the gene, their children will have to care for them and they wish to spare their children that experience. 'So it's not forty carefree years,' she says.

The moral issue becomes even more acute when one considers the polygenic diseases such as heart disease, diabetes, cancer and schizophrenia. It is perhaps misleading to lump them together for they are under varying degrees of genetic 'control'. According to Sir David Weatherall, 'it is very likely that some people are predestined to get diabetes'. Late-onset, or non-insulin dependent (type II) diabetes seems to result from the action of several genes, but it is 'very strongly genetic' he says. In identical twins, if one gets it, then it is overwhelmingly likely that the other will suffer from type II diabetes as well. With heart disease, on the other hand, there is only a 20 per cent incidence of a second twin succumbing to the disease when the sibling has it, suggesting a much weaker genetic connection. Most cases of cancer seem to be the result of damage to genes sustained in adulthood and thus to be a consequence of the interplay between genes and the environment – smoking cigarettes and being exposed to X-rays and other radiation are two of the known environmental causes of cancer.

But suppose that the genes controlling the onset of diabetes were to be isolated and a characteristic pattern were to emerge which had 100 per cent predictive power. One consequence would be the development of tests to pick out those who are predestined to get the disease. A prenatal test would no doubt eventually be possible. Would it be morally permissible? The child would have to live with the foreknowledge of future disease even if, by choice of lifestyle, they might be able to avoid or minimize the disease. But if prevention is not possible, ignorance would have been preferable. A child brought up with such knowledge might well come to hate its parents for bringing it into the world with this burden. So would it therefore be morally acceptable to abort a foetus which had tested positively for late-onset diabetes? The very act of conducting the test might well condition what action is to be taken in the event of a positive diagnosis, not because the predicted disease is so severely disabling (it is not) but because the knowledge of one's predestination is.

What then happens for late-onset polygenic diseases where the outcome cannot be predicted with 100 per cent certainty? Suppose, at some point in the future, there were a prenatal test for propensity to heart disease but that the results of the test were only probabilities. If parents are justified in terminating the pregnancy of a foetus with 98 per cent chance of getting Huntington's late in life, at what level of probability are they justified in terminating a pregnancy where the foetus has a chance of getting heart disease – 80 per cent; 60; 40?

The availability of tests that reveal inherited susceptibility to conditions such as cancer or heart disease do not just pose moral problems for prospective parents. Such tests will make more acute the public policy problems for countries with private health care systems. It may be possible for someone with an inherited predisposition to cancer to change the environmental contribution and so avoid or at least delay disease development. They could stop smoking and change their diet, for example. This would clearly be good news for the individual. But consider the situation from the point of view of the providers of health care and suppose, for a moment, that no such preventative action is possible.

In a socialized health care system this simply does not matter. Everyone in the country is covered by a health service, which is set up to deal with the average national death rate from cancer. The availability of genetic testing would not alter the actuarial figures for the annual death rate (on the assumption that no preventative action can be taken) and so there would be no difference in the cost to the system. Indeed, in this model, there is no advantage in anyone knowing their genetic susceptibility (and probably much personal psychological disadvantage) while offering the tests would itself represent a cost. It is likely therefore that a socialized health system would not use such genetic tests, unless they promised some benefit in terms of better health and longer life to the individual tested. In contrast, a private health system based on medical insurance is likely to use genetic testing because of the benefits which accrue to the insurers rather than to the individual being tested. It is obviously in the interests of a medical insurance company to be able to guarantee that those it insures are not going to require expensive treatment for cancer at an early age.

But before the British or others gloat over the obvious deficiencies of the US private health system, it should be remembered that health provision is not the only service provided through insurance. Old age pensions are often funded by life assurance. My profession of journalism is notoriously fluid, and few journalists stay long enough with one employer to benefit from joining a company pension scheme. To provide for my old age, and for my wife's, I recently started a personal pension plan backed by a life assurance policy. The application form contained a clause which gives the life assurance company authorization to seek 'medical information from any doctor who at any time has attended me concerning anything which affects my physical or mental health'. The company assures me that under the Access to Medical Reports Act, passed by the UK Parliament in 1988, I have greater control over the use of my personal medical records and additional rights as a consumer. Among them is the right to stop the assurance company seeing my records. But if I exercise that right, I will not be able to join the pension scheme and so provide for my old age. As and when genetic tests become available, should my insurers have right of access to the results? As society already accepts that insurers can have access to medical records, it seems to some observers but a trivial extension of an already accepted principle to allow access to the results of genetic tests.

But allowing health insurers access to genetic records may be to the general disadvantage. Individuals may well feel that they can do nothing about their genetic constitution – they are born with it, they did not choose it, and they cannot alter it – so they should not be penalized for matters over which they have no personal control.

In a society which respects the right to privacy and to private property, could anything be more personally private than one's genetic constitution? Therefore, one might say, information about a person's genetic constitution should remain privy to that person only. But the issue is not clear cut. For a start, a person's spouse and children almost certainly would have rights to know. More generally, there is the matter of equity between the individual and his insurers. All insurance is a bet. If I know I have a genetic predisposition to heart disease, but the insurance company does not know, I am in a position to change the odds of the bet: I can take out much more insurance. As Wexler comments, 'Anything that is supported through life insurance is a major problem. If people know they are at risk they could buy enormous insurance and skew the statistics.' On the other hand, the company might try to skew the statistics by refusing to cover those diagnosed to be at genetic risk. In such an event, Wexler believes, the only way to get cover would be to get a job which carries

Insurance

with it an occupational pension where there is no genetic screening.

Concern in the United States has focused on the possibility that the workplace might become the venue for genetic screening. Employers potentially have two commercial interests in screening their workers for genetic traits. Some individuals, those with G6PD deficiency for example, will become ill if exposed to certain chemicals which might customarily be present in the workplace, so if those at risk are identified by screening they can be assigned jobs where they are not exposed. The second interest is very different and again relates to the system of private health care insurance. Most people in the USA depend on their employers for health care provision. The cost of health care insurance premiums has risen dramatically in recent years and many large companies are adopting self-insurance plans. Large employers assume the risk for the health care of their employees directly instead of purchasing insurance. They could therefore be interested in screening workers and job applicants to exclude those who are likely to develop expensive diseases. The motive here is very different from the putative case of G6PD. The concern would be simply to exclude those at risk of developing a disease later in life which would increase the company's health care costs.

To assess the extent of genetic screening and monitoring in the workplace, the US Congressional Office of Technology Assessment carried out a survey in 1989 of the *Fortune* top 500 companies in the USA, together with the fifty largest utilities, thirty-three trade unions, and a sample of 1,000 other companies each with a workforce of more than 1,000. The survey updated an earlier one carried out in 1982. It found surprisingly little or no increase in the number of companies conducting genetic screening or monitoring in the workplace. None of the companies surveyed anticipated using direct DNA analysis in the five-year period following the survey. However, it may be that this concern over workplace testing is premature for two reasons, one to do with the technology, and the other related to the science of genetic diagnosis. The technology for taking and analysing DNA samples cheaply and effectively did not exist at the time of the survey. In addition, there is only a limited range of conditions for which such tests currently exist. The commoner diseases which have a genetic component, such as heart disease and cancer, are complex polygenic conditions whose underlying genetic basis has yet to be unravelled. Consequently, the development of a predictive test based on genetic analysis lies a long way in the future. It remains to be seen if companies will maintain their current lack of interest in genetic testing if and when cheap tests become available for more common diseases.

Genetic discrimination of a sort already exists in the workplace. The inherited defect of colour blindness is an interesting example, because it was recognized long before the modern methods of molecular analysis were invented. Some jobs are forbidden to those with this defect: no one who is colour blind can be a commercial airline pilot. Superficially, this appears good evidence for saying that someone's genetic constitution cannot be their own confidential property – it is, after all, in the public interest that aeroplanes should be safely piloted and the inability to distinguish red and green could be a serious hindrance. But the point here is not that a man who carries the gene is barred from a particular employment, the disbarment arises because the defect carried by that gene is actually expressed – it is the actual disability which leads to the ban. The airline is interested in the effect, not the cause, and while the effect may betray part of the man's genetic constitution, that is not relevant information.

Genetic discrimination such as this is accepted for two reasons. One is that the disability affects the safety of others. The second is quite simply that such examples are exceedingly rare, because our knowledge of genetic defects is comparatively poor. However, knowledge will accumulate dramatically as a result of the human genome programme and the opportunities for more refined decisions on the genetic selection of workers will increase enormously in the future.

Consider the case of genetic predisposition to cancer. It could be in the interests of the nuclear power industry, say, to screen prospective employees to ensure that only those who are less susceptible than the norm are taken on as workers in radiation areas. One could imagine a less scrupulous company than those currently operating in Britain deciding, either implicitly or explicitly, that, as its screening programme had selected only those employees with below average cancer risk, it need not spend too much of the shareholders' money in keeping radiation levels low. Although both genes and the environment contribute to cancer, society has hitherto been able to alter only one of these variables – the amount of radiation in the workplace. But genetic testing will open up the possibility of altering the first variable

– the genetic structure of the workforce – by selecting only those individuals with the desired genetic constitution. It would then be less imperative to reduce radiation dose to the workers because, thanks to their genetic profile, they would be better able than the average to resist its carcinogenic effects.

In practice, this is unlikely to come to pass. Public concern about nuclear power is already so great that neither government regulatory bodies nor public opinion would permit such a trade-off. But such a situation could arise in more ambiguous circumstances. And whatever the details, the question has to be asked whether there is indeed anything necessarily shocking about such an attempt. Who is to say where the balance lies in cancer prevention?

The ability to change genetic factors where before only environmental influences could be controlled will have an impact in other areas of life. One of the most sensitive could be in education and intelligence.

Some wealthy British families consider it desirable to send their sons to Eton. Even now, parents must put the boy's name down almost at birth. If rising academic standards force up the competition for places, they may discover that this is no longer enough and that they have to submit the genetic profile of the foetus weeks after its conception, as a way of certifying that the future Etonian will be highly intelligent. Other parents, in the future, may try selectively to abort foetuses until they conceive one which prenatal genetic tests indicate has the right combination of intelligence genes. Dr Robert Goodman, of the Institute of Psychiatry in London, believes that within five to ten years it may be possible to detect characteristic patterns in human DNA correlated with higher intelligence. Other scientists believe that this underestimates the difficulties of correlating genes and intelligence, but there is such obvious scope for abuse that there is a need for public discussion of the ethical and moral issues before the power of genetics is harnessed to predicting a foetus's future intelligence.

Goodman fears that if prenatal diagnosis is eventually able to pinpoint genes associated with intelligence, there will be 'substantial capacity for abuse'. One can imagine, he says, some parents using selective abortion until they achieve a child with the desired genetic profile associated with high intelligence or even opting for *in vitro* fertilization and embryo screening to implant only the one with the desired characteristics. Prenatal diagnosis of the sex of an embryo has already been abused in this way – usually to abort female embryos. It would not matter, he believes, if the procedures were expensive or even if the genetic prediction accounted for only 20 per cent of the variance in intelligence. 'One has only to look at how much some parents are willing to spend on private schools in the hope of increasing their children's educational attainments.' It may be possible, he says, for them to attain the same end at lower cost through selective abortion guided even by relatively imprecise genetic predictions.

Intelligence results from many genes acting together and from their interaction with the environment. Goodman says that while many genes contribute to intelligence there could be a small number of genes each accounting for, say, about 10 per cent of the variance in intelligence. If this is the case, then he is confident that they can be located within ten years. However, if 1,000 genes are involved, each contributing only 0.1 per cent, 'there is no hope of finding them all'. Goodman believes that the real situation is probably somewhere in betweeen these two extremes, with very many genes being involved, of which a few are major players.

Research into intelligence and genetics could have beneficial effects. By analogy with the case of PKU, it might be possible to identify factors in the diet of those who have learning difficulties such that dietary modification could ameliorate their mental handicap. But because of the temptation to misapply any new knowledge, society must decide whether it needs to set rules to prevent researchers even investigating certain avenues.

Selecting your child's characteristics need not be confined to intelligence. Some parents might engage in selective abortion for frivolous reasons – to secure offspring with blue rather than brown eyes, for example. From today's perspective, a society in which such actions were commonplace – or even in which they were permitted but occurred rarely – would be deeply unattractive. It seems unlikely that many couples would avail themselves of the technology, for it would introduce so calculating and mechanistic an approach to the business of reproduction as to be aesthetically displeasing. But however unaesthetic, it is not easy to show that such practices stretch morality beyond the boundaries of what is considered acceptable today. Nor is it easy to work out how such practices could effectively be forbidden.

Consider for a moment the position of the parents and of the child. Having achieved a child with the genetic characteristics they desire, the parents can be expected to cherish and look after the child with greater attention and devotion than they might otherwise have done. Similarly, the child will go through life with the reassurance that he or she was not the consequence of blind chance – the lucky draw from the genetic lottery – but that his or her parents went to considerable trouble (and expense) to bring about his or her birth.

It is hard to see how anyone has been harmed or damaged by this procedure: on the contrary, everyone seems to have benefited. The only 'people' to have been harmed are the foetuses who were not brought to term. Do they have any claim to moral consideration? In most Western countries nowadays, there are few people who would object to the termination of a pregnancy if the alternative were the birth of a child with a fatal disease – Tay-Sachs for example – or even a crippling disability. The justification is that such a child would not benefit from a worthwhile life, but that its parents may later be able to conceive a child who can have a worthwhile life. The question is then the definition of a 'worthwhile life'. In industrial societies, it is clearly advantageous to be of above average intelligence – life is more worthwhile – so why should there be any legal block to parents seeking to achieve this end for their children, if the technology permits? At the most extreme, it could be argued that if the technology for such genetic screening were available but society forbade its use, then some parents might be resentful of their unchosen children who would therefore not have a happy and fulfilling life – although one imagines that these would be very few indeed.

It is difficult in fact to see how such practices could be prevented. Couples already undergo genetic testing and prenatal screening for known genetic diseases. Is their clinician of the future to give them only some and not all of the results of the tests carried out on the foetus? The foetus's genetic profile is the consequence of the parents' genes: the doctor is merely the skilled middleman who interprets the genetic profile for the parents. If there is any right to ownership of genetic information, it must surely rest with the couple, not with their doctor.

The discussion of intelligence has so far focused solely on the motivations and actions of individuals as prospective parents. Society might also have an interest on the grounds that it would be a good thing generally for there to be lots more intelligent people around. There are two dangers here. One is the elementary point that just as people are more than their genes, so there is more to human character than intelligence. The 'value' of another human being, both their moral value and their personal or social worth, depends on much more than just their intellectual capacity. Intelligence tends in any case to be relatively narrowly defined, usually in relation to scholastic ability. In terms of the contribution that an individual might make to society as a whole, other traits are equally, if not more important. Few captains of industry, for example, have been noted intellectuals nor have successive US presidents shone in terms of their scholastic ability. One of the most intelligent of presidents in recent times, Jimmy Carter, was one of the least successful. Human beings are social creatures and the communities in which we live depend for their smooth functioning on the quality of the interactions which individuals have with each other. Thus human insight and understanding, a sense of what things are right and wrong, what is acceptable behaviour for the group in which one lives, are all necessary qualities, and do not depend on a greater than average level of intelligence. There is no point in breeding a generation of *Übermenschen* if they behave to the less favoured with arrogant cruelty born of their consciousness of how favoured they are. Breeding superintelligent moral cripples would not be an advance.

The second objection to any involvement of wider society in such an issue is founded on the question of who should control such technology and decide which children would be born more intelligent than the norm. Except in a few communities, such as the New York Hasidic Jews mentioned earlier, most people believe that decisions on whether or not to have children are personal and that wider society, much less the state, should not determine such choices directly. Society and the state may obviously influence someone's decision in many ways, but that is different from dictating what the decision should be. To allow the state or the community into this area of private life for the purposes of determining the intelligence levels of one's children must surely represent an intrusion which most would wish to resist. Yet the technology to achieve selective breeding will be expensive and rationing will be inevitable. If decisions are left to individual couples, then rationing will be on the basis of price and those who are wealthy are likely to have preferential access.

Inequitable though this would be, it might be preferable to rationing imposed by the state.

This chapter has travelled a long way from the malarial parasite and the genetic disease that it has brought about. But the distance is more apparent then real, for the same issues underlie testing for most genetic diseases. The first question is the rationale for the test itself – what good does it do to know that a person carries a genetic defect? The second is access to that knowledge – who has the right to know that a person carries a genetic defect? For the time being, the first application of the new knowledge generated by the Human Genome Project will be in devising genetic tests. Cheap and effective gene therapy, as discussed in the next chapter, is still a long way off. Diagnosis without hope of cure is likely to be the norm for many years to come. The main application of the new genetics is likely to be in prenatal diagnosis and, for some couples, will result in early termination of an affected pregnancy. Benefit from the new genetics is likely to be surprisingly small. Indeed, genetic diagnosis is likely to bring with it significant disbenefits, including the risk of social stigma, and of discrimination at work or in the insurance market if the results of genetic tests leak out from the privacy of the doctor's consulting room. Yet third parties are always interested in the results of genetic tests, as shown by the case of testing for sickle-cell anaemia, and some outsiders may have rights to such knowledge. I do not own my own genes; I was 'given' them by my parents and I have passed them on (or, at least, half of them) to my children. My immediate family have a legitimate interest in knowing my genetic constitution, so I cannot claim it as my private property. But the most important question to be posed in the short term by the Human Genome Project, a question which society has not even begun to answer, is who else has the right to know what is written in my genes and who can constrain how I act on that knowledge?

7

New Genes for Old

IT is the stuff of the Hollywood horror movie: the genetic surgeon stands poised ready to graft new genes into his patient, to transmute the recumbent figure on the operating table into some new species of superbeing or some awful mixture of human and beast. Of all the developments associated with human genetics, gene therapy most arouses fears of a latter-day Dr Frankenstein creating new and monstrous forms of life. In fact the outlook for gene therapy is currently very limited. Replacing a defective gene is one of two logically distinct ways of dealing with genetic disease; the other is to replace the deficient protein. These procedures have their attendant moral problems, although some of the issues raised are common to medical practice in general and are not unique to the science of genetics.

None the less, there are new worries. Concern has focused on what would happen if gene replacement were carried out not to cure disease but to enhance certain characteristics in an otherwise normal person. As this chapter will show, such dreams seem likely to remain science fiction for many decades. A further concern is with the effects of gene replacements that can be passed on to future generations, thus permanently altering the human genetic legacy.

But the emphasis on genes forgets that a gene is merely a convenient route to obtaining a protein: it is proteins which make human beings what they are. In contrast to the emphasis on genetic engineering, few people have worried about the possibility of altering a person's characteristics by giving them additional doses of the proteins corresponding to certain genes. One of the first things that is done even now when a human gene is isolated, is to pop the gene into bacteria or other cells and get them to produce the corresponding protein. Unlikely though it may seem, the way we live our lives is far more likely to be affected by the splicing of human genes into bacteria to manufacture human proteins to be used as pharmaceuticals than by the splicing of those genes into humans.

The genetic engineering industry is not directly relevant to the tale of how human genes are mapped and sequenced, but it will play a major part in the consequences of the Human Genome Project. Looking back on the history of a couple of the biotechnology industry's products amply illustrates both how modern genetics promises to alleviate human suffering and how it may open up new moral dilemmas. Both products are human proteins which were once scarce but are now (comparatively) plentiful: human growth hormone, whose story was touched on in Chapter 4, and another hormone, erythropoetin, which acts to increase the body's supply of red blood cells. One of the main outcomes of the Human Genome Project will be the isolation of tens of thousands of human genes from which the corresponding proteins can be mass-produced. Neither human growth hormone nor erythropoetin was a consequence of the Human Genome Project as such but they anticipate things to come. If society has handled the sudden availability of these proteins well, then one may be confident that it will be able to handle the avalanche of new materials; if these substances have been abused, then what lessons can be drawn to prevent a repetition?

There are about 3,000 children in Britain whose bodies do not make enough of the protein, human growth hormone, to enable normal growth. Only very rarely is the condition due to an inherited defect; in some cases, there seems to be an association with complications during birth itself, whereas in others the cause is simply not known. Between 1959 and 1985, such children were treated with injections of human growth hormone extracted from the pituitary glands of human cadavers undergoing post-mortem examination. This was the only source of the hormone. Tragically, unknown to those who were extracting the hormone from the deceased human 'donors', it now appears that at least one of the donors must have been developing a rare form of brain degeneration, known as Creutzfeld–Jacob disease (CJD). This disease appears to be similar to scrapie in sheep and to 'mad cow disease' (BSE). The protocols for preparing the pituitary extract precluded the use of tissue from anyone who had died from or was suffering from a neurological disease but the signs of CJD were not detectable at the time.

The disease can be transmitted by injection of contaminated human brain tissue and it appears that the infectious agent was transmitted with the human growth hormone to children being treated for stunted growth. At least eight of the children treated in this way have died in Britain and the pattern of contamination has been reported in other countries, including the USA, France and Australia: more than twenty people around the world have since died from CJD. Since 1985, however, doctors have been able to prescribe human growth hormone made by genetic engineering. (The process was described in Chapter 4.) One of the first companies into the field actually constructed part of the gene synthetically in the laboratory, using the chemical scissors of restriction enzymes to snip out the remainder, transferred the gene into bacteria, and grew them in large numbers in culture. In this way, it is possible to produce large quantities of the hormone without any risk of CJD contamination whatsoever. All human growth hormone used in treatment today is manufactured in this way.

But the plentiful supply of safe human growth hormone raises unexpected ethical questions, for it is now also being used to treat children who are short but who do not suffer from pituitary deficiency. The availability of the virtually risk-free drug has brought about a 300 per cent increase in the number of children being treated in Britain since 1985. The drug is still not cheap – one week's supply for treating a child with dwarfism costs about £150 and to increase a child's height by about an inch might require a programme of injections costing more than £7,500 – so price is likely to be a limiting factor in its wider use. But the availability of the drug is blurring the distinction between medication being given as a 'remedy' to cure a defined disease and as a 'cosmetic'. In general, it is an advantage in our society to be tall – we suffer from 'heightism' as well as racism, sexism and ageism. If human growth hormone is now available, why should not a fond parent seek to give their child an advantage in life – even though there is no medical reason for the child to receive the drug? There are rumours of doctors in the USA seeking to provide their own normal-height children with just this advantage of extra height conferred by the genetically engineered hormone. Moral questions concerning access to the fruits of genetics research will become more numerous and more acute as the Humane Genome Project proceeds and the misuse, if it is a misuse, of pharmaceutical proteins, which are indistinguishable from the real thing because they are copies of natural human protein, will be one of the first issues that society has to cope with.

In fact, the problem is already here. In June 1991, an article in the *New York Times* magazine drew attention to how Genentech, one of the leading manufacturers of human growth hormone, was promoting the drug for use in children who were short but not hormone deficient, even though the clinical evidence as to whether extra human growth hormone really does make 'normally' short children grow into taller adults was ambiguous. Genentech had invested heavily in biotechnology, the article noted, and so the company had a considerable incentive to expand the market for its human growth hormone to try to recoup its very considerable research and development costs. 'Genentech began avidly pursuing the largest group of potential users available to it,' the article continued. What the children all had in common, it was stated, was being in the lowest three percentiles in height. But this is a statistical measure which reflects the actual distribution of children's heights. One might as well worry about the fact that around half of any given sample of children will be below average height – the definition of 'average height' in itself ensures that roughly half the population will be above average and the other half will be below the average. About 3 million children are born every year in the USA and some 90,000 of them will be below the third percentile for height – by definition. If all were treated with human growth hormone, the market for the drug would top $8 billion. But as an article published in 1990 in the *Journal of the American Medical Association* pointed out, the use of the bottom three percentiles as the measure of shortness has a built-in escalator: even if everyone's height were increased by a couple of inches, there will always be a lowest three percentiles. The average height may have increased by two inches but someone will still be shorter than the average.

The burden of the *New York Times* article was that the abundant availability of human growth hormone was changing the definition of disease. What was in times of scarcity a relatively clear-cut criterion of dwarfism due to human growth hormone deficiency has now become more diffuse, until the simple fact of being that much shorter than the average makes someone a suitable case for treatment. This demand is fuelled by existing discrimination against those who are short, and in a free-market economy it is in the commercial interests of the manufacturers of the drug to meet that demand. The biotechnology industry is therefore providing new options for people. Previously, people who were below average height had little option but to come to terms with their condition and get on with their lives. Now, they can turn to hormone therapy to change it. Commercial pressures are setting a question to which there is no satisfactory answer: what is normal? The human population is wonderfully diverse and varied, with a distribution of characteristics and traits which spreads widely on both sides of 'the average'. But the lesson of human growth hormone seems to be that, in the case of height, those who are below average height do not always regard this as a welcome manifestation of human diversity but are willing to pay large sums of money for long-term treatments simply to come up closer to the average. Some may be willing to do so even if the efficacy of the treatment is doubtful. The results of a drug's clinical trials, especially for a treatment of this duration and imprecisely defined outcome, are seldom clear cut and unambiguous. Those who are driven by deep unhappiness about their 'condition' will read hope into ambiguity and demand a treatment which carries even a faint prospect of ameliorating their state.

Nor are the redefinitions of disease brought about by human growth hormone confined to shortness. There is some controversial evidence that if elderly men undergo a course of injections, this can reverse a substantial proportion of muscle shrinkage and accumulation of fat that goes with old age. This correlates with existing knowledge that as people get older, their rate of natural human growth hormone production slows. As the *New York Times* article concludes, 'the question was whether giving it [human growth hormone] to old men, 40 or more years after they had reached their final height, wasn't simply more unwarranted tampering with nature. Short kids made taller, old people younger – appealing notions perhaps, but where does it all stop? Is ageing too, like shortness, now to be considered a disorder?'

In addition to its official medical use, human growth hormone is being traded on the black market and used by athletes artificially to boost their sporting prowess. There is definite evidence of its illegal use among body builders and professional players of American football and the likelihood is that genetically engineered human growth hormone is being used in other professional sports as well. Athletes believe that the hormone builds up their muscles, and strengthens tendons and ligaments. According to the scientific journal *Bio/technology*, human growth hormone is replacing anabolic steroids as the preferred clandestine performance enhancer. Because the hormone is indistinguishable from the real thing, it will not show up on the tests now routinely administered to athletes participating in major competitions. The journal quotes Lyle Alzado, a former member of an American football team, the Los Angeles Raiders, describing human growth hormone: 'It's the drug of choice. I took growth hormone for 16 weeks. I went through tests from the NFL [National Football League] and I didn't get detected for it.' The use of human growth hormone has been banned by the NFL, the US Olympic Committee, and the US National Collegiate Athletic Association.

However, there is some question as to whether human growth hormone does actually increase performance. Although it stimulates protein synthesis, there is no guarantee that it will specifically increase the growth of the required muscles. One study of the effect of human growth hormone on weightlifters found that although those taking the hormone gained weight faster than a similar group taking a placebo, the weight gain was probably due to water retention and to an increase in lean tissue other than muscle: bone and connective tissue, liver, spleen and kidney. Such a result is slightly surprising since, in animal husbandry, the use of the animal growth hormones is being promoted precisely to increase the amount of lean muscle tissue in beef cattle and in pigs, and it seems unlikely that the analogous hormone would have different effects in humans. Whatever the long-term effect, so great is the demand for the hormone among athletes that some enterprising dealers have taken to selling counterfeit human growth hormone.

While human growth hormone may not do athletes any good, it also probably does not do them much harm. The same cannot be said for one of the other human proteins made by genetic engineering and now circulating on the athletics black market. Erythropoetin (EPO) stimulates the bone marrow to make red blood cells and is used in the treatment of chronic anaemia. It may also have contributed to the deaths of as many as eighteen European professional cyclists within a period of four years. At least one incident has become public knowledge: Lisa Draaijer, the widow of a twenty-seven-year-old Dutch cyclist Johannes Draaijer who died in his sleep in 1990, believes that EPO may have been a factor in her husband's heart failure and is endeavouring to discourage the use of EPO in sport. Athletes in endurance sports such as cycling turn to EPO because the drug

139

increases the number of red blood cells and so enhances the oxygen-carrying capacity of the blood. Unlike human growth hormone, EPO does seem to deliver the desired effect. In an investigation of abuse of the drug, *Bio/technology* reported on a study which had shown a 10 per cent increase in performance among men taking EPO. But there are problems: increasing the concentration of red cells in the blood can lead to blood clots and can put too great a strain on the cardiovascular system. The effect is exacerbated by the very endurance sports in which the athletes taking EPO engage: sweating and dehydration concentrate the blood still further. The blood turns from liquid to a sludge and, because red cells persist for up to 120 days, the strain on the cardiovascular system persists long after the event for which the athlete was trying to boost his performance. Again, the drug is next to impossible to detect, because it is human EPO, but none the less the International Olympic Committee has added it to its list of banned substances.

One reaction to these case studies of human growth hormone and EPO might be 'So what?' Athletes have been injecting themselves with noxious compounds for many years in order to gain an illicit advantage over their opponents and some of the compounds in common use – anabolic steroids for example – are at least as damaging if not more so in their effects as either of the two new drugs. The manufacturers of the drugs have moved swiftly and responsibly to try and prevent 'leakage' of their products on to the black market and to persuade athletes of the perils of using such compounds. Ultimately the sports market is small and does not represent widespread abuse which would cause problems for society as a whole. But the interest of these two drugs is that they are examples of what we may expect when human genes are isolated and put to work within bacteria (or yeasts, or whatever) and the corresponding proteins produced in large quantities.

Consider a hypothetical example. One result of the genome project will be to identify and isolate genes and proteins that function in human brain cells. Suppose that subsequent research discovers that one of these proteins increases the speed with which nerve cells transmit their signals. Furthermore, laboratory mice injected with the corresponding murine protein solve puzzles much faster than normal mice. Finally, it is established that people with high IQs have higher concentrations of this protein in their brains than people with below-average IQs. Any commercial company which had the patent over such an alleged IQ-enhancing protein would command a market of billions of dollars. This would still be true even if the efficacy of the protein were doubtful. For what is known of how the human brain actually works suggests it is unlikely that increasing the speed of nerve conduction within the human brain would do anything to enhance intelligence, or any other property for that matter. It is overwhelmingly likely that the role of genes in the development of intelligence is to determine the connectivity of one neuron to another in the brain and that this is finalized early on, so that taking a drug as an intelligence enhancer is unlikely to have much effect. Intelligence (which is not the same thing as IQ) is clearly a complex piece of human behaviour partly influenced by a person's genetic make-up and partly by the extent to which they have enjoyed a stimulating environment which gives their brain a chance of expressing its full potential. In addition, many genes have to work together in concert to produce the ability which we know as intelligence: there is clearly no single gene for intelligence, so the idea that an excess of one gene-product could enhance intelligence is clearly misguided. But in the public ferment that would follow the discovery of an alleged IQ-enhancing drug, these points would be sidelined as footling academic nit-picking. The fact that there is no single gene for athletic prowess does not stop athletes from buying human gene products on the black market – even though the consequences may be fatal.

Despite these cautions about the nature of the brain, some scientists have actively promoted the idea that speed of nerve conduction is related to degree of intelligence. Arthur Jensen, an educational psychologist at the University of California at Berkeley, is one who believes that there is a direct correlation between the speed at which nerve impulses travel and level of intelligence. Professor Jensen first came to public prominence in the late 1960s for his highly contentious views that IQ was predominantly determined by genetic inheritance, and that since blacks had scored worse on a series of IQ tests than whites, the average intelligence of blacks was lower than that of whites. The inference was that the programmes inspired by the 1960s civil rights movement to better the lot of the American blacks were a waste of time and money because the inferiority of the blacks was genetically pre-ordained. Jensen's conclusions are dismissed by professional geneticists who believe that he pushed the connection between genes and intelligence too far, that his ideas are too simplistic to be adequate for so complex a human attribute as intelligence, that he has misunderstood the significance of the statistics, and that his interpretation of the data was wrong.

None the less, by 1991, Jensen was testing the speed at which the electrical signals generated on the optic nerve travel through the brains of students participating in his experiments and once again interpreting the results in terms of racial categories. A picture is flashed up in front of the participant, the optic nerve 'fires' and a fraction of a second later corresponding electrical activity can be detected at the back of the head. Jensen used the external dimensions of the participant's head to gauge the distance the nerve impulse has to travel and, from the time delay, worked out the speed. He claims to have detected significant differences between whites (who are quicker) and blacks (whose nerves run more slowly). The crudity of these measurements and of the underlying hypothesis cannot be emphasized enough, and it must be stressed that the whole enterprise is far from the mainstream of scientific research. But research of this nature is being pursued by a senior academic at one of the world's most respected universities and this leads some people into believing there must be something in it.

The suggestion that some chemical compounds might boost IQ received extraordinary exposure in Britain in 1991, when it was alleged that vitamin pills taken as supplements to the diet could increase children's performance. The research – with which a vitamin supplement manufacturer was closely involved – was widely reported in the media in an uncritical fashion. Indeed the claims were virtually promoted by a BBC television programme (despite the fact that the BBC is not supposed to show any favour to commercial companies) and were endorsed by several scientists, including the professor emeritus of psychology at University College, London, Hans Eysenck. A scientific journal of which Eysenck is editor published a special issue carrying the results of the research to coincide with the BBC TV programme. Within twenty-four hours, virtually all shops had sold out of vitamin pills. Yet among most researchers professionally involved in nutrition there is some scepticism about the claim that vitamins can enhance intelligence.

There is little sign that anyone in the scientific community or elsewhere is facing up to the possibility of similar developments arising from the Human Genome Project. Nor would the commercial gains be confined to IQ enhancement. It is possible to imagine that if a protein were isolated which, when given to children, ensured that they grew into adults of lighter skin colour than they would without the drug, it might find a market among the Afro-Caribbean population living in Europe and North America. (Without the benefit of genetic techniques, pop singer Michael Jackson must have spent a considerable fraction of his fortune on plastic surgery to convert his appearance from the negroid features with which his genes endowed him to the simulacrum of a white man.) Other products, including, for example, drugs to boost the action of the immune system, to reduce one's likelihood of developing cancer or heart disease, could also find an extensive market. Most certain of commercial success would be claims that particular preparations could enhance human longevity. If there are about 100,000 genes to be discovered – even if there are only 50,000 – the potential for exploiting human frailty is enormous. The Human Genome Project carries with it the possibility of spawning a new generation of snake-oil salesmen – many of them, on past form, claiming to be scientists themselves and wrapping their claims in the language and the respectability of science.

Problems may stem from honest scientific errors. It is worth considering an example from a completely different area of science: cold nuclear fusion. In 1990, two respected chemists made the astonishing claim that they had succeeded in fusing atoms of heavy hydrogen in a controlled manner using simple laboratory benchtop apparatus. Fusion (in an uncontrolled manner) is responsible for the explosive power of the hydrogen bomb and is the energy source which drives the sun. Physicists around the world have been trying for several decades, with the expenditure of hundreds of millions of pounds, to control the process using enormous experimental installations. The suggestion, by Stanley Pons and Martin Fleischmann, that they had succeeded where everyone else had failed and, furthermore, that they had done so with cheap and inexpensive apparatus generated extraordinary excitement. Very quickly, however, the scientific community became convinced that the two chemists had not produced controlled nuclear fusion and that the energy which was apparently being produced was simply an artefact of the apparatus. The majority of scientists who investigated the phenomenon became convinced that Pons and

Fleischmann had made some honest mistakes. However, even after all the contrary evidence, there remains a small group of committed advocates of cold nuclear fusion, with Professors Pons and Fleischmann among them.

The scientific community is ill-equipped to deal with such eventualities. Few scientists properly understand the economic and commercial institutions which translate discoveries in the laboratory into products which are marketed for profit by the reputable pharmaceutical industry, let alone by fringe practitioners. They have little idea of how a scientific discovery can become a patentable piece of intellectual property and of how the pressures of the market-place may influence the application of their discovery. How are lay members of the public to react if they are confronted with dissent and conflicting claims from within the scientific community as to the efficacy and social utility of a particular compound?

Two situations could arise. In one, there may be genuine scientific uncertainty – the supplementary uses of human growth hormone may be a case in point – where clinical trials have either not been completed or have yielded ambiguous information which different scientists interpret differently according to their own background and prejudices. This is a case of genuine scientific debate, which it would be in nobody's interest to stifle. But the other case is where people at the fringes of the scientific enterprise make large claims on areas of scientific territory where they have no real competence – the genetics of intelligence may well be one example for the future – as a result of the Human Genome Project. Here the debate may not be genuine, despite the fact that some of those making such claims have respectable academic qualifications. Such spurious debate confuses, and furthers no public interest. I do not wish to suggest that a form of scientific censorship may be needed, but the scientific community will need to devise a mechanism by which it can swiftly and very publicly rebut claims that have no scientific foundation. This is not something which it has shown itself ready or equipped to do in the past.

Formal procedures already exist to control and regulate the introduction of a new drug on to the market. Consider again the hypothetical example of an IQ enhancer. There are three aspects to the availability of any drug. Is it safe? Does it work? Is it morally and socially acceptable that such a compound should be available, either as a prescription drug or even sold over the counter? The first two points are technical ones, and it is to these aspects that government regulators – the Food and Drug Administration in the USA, the Department of Health's Committee on the Safety of Medicines in the UK, and so on – address themselves. The organizations which regulate the safety of new drugs tend to confine themselves to technical questions, they do not explicitly consider the social or moral consequences of the marketing of specific products. Even so, although questions of safety and efficacy may be technical, they are not easy to answer. Since the thalidomide tragedy in the 1960s, testing of new drugs has been tightened up still further, but there is still no absolute guarantee that compounds which have tested safe on animals and in laboratory analysis will be safe when administered over long periods to humans. This would be particularly true of a compound expected to affect the workings of the brain: it might change personality, induce manic behaviour or produce some subtle long-term effect which might show up only when large numbers of humans have taken it over a period of many years. In principle, however, a clear-cut answer can be given to the question of safety.

Judging the compound's efficacy would be rather more tricky. Rats displaying enhanced intelligence by solving the problems of a laboratory maze would not really be enough. Other questions would have to be asked with respect to human use: how high a dose produces an observable effect on performance in IQ tests? Would a child have to take the drug continually? If they stopped, would the nerves atrophy, leaving the subjects less intelligent than they would have been had they never taken the compound in the first place? An underlying question here is, if the drug were to be licensed, what disease is it supposed to be curing?

Perhaps the first use would be for those who were so severely mentally retarded as to be unable to look after themselves. If they demonstrated a significant improvement after receiving the drug, it would naturally be in the public interest for the drug to be licensed because such people might then be able to leave the (usually state-run) homes in which they had been cared for and thus be less of a financial drain upon the public purse. This use would be medically prescribed as a treatment for a recognizable condition, but would it be in the public interest for the drug to be more widely available? In most countries, once a drug has been licensed, even though it may be recommended for only a few conditions, decisions on its use are left

145

to the discretion of medical practitioners who may choose to broaden the range of conditions for which they prescribe the drug.

At first sight, the reaction might be to say that injecting a drug into normal people to boost an already adequate IQ is an abuse and should not be allowed, even if the drug has passed the tests of safety and efficacy. But is it in fact an abuse? Certainly, the availability of such a drug would have much the same effect as the *New York Times* descried with the availability of human growth hormone. The definition of disease – pituitary dwarfism – is subtly changing to below-average height. Similarly one might expect an IQ-boosting drug to shift the emphasis from recognizable mental deficiency to below-average IQ. In a society such as ours which values intelligence so much, is there anything realistically that can be done to prevent this? Even were it possible, should anything be done to prevent it?

One argument against the wide availability of such a drug might be that if the drug is expensive, the existing inequity in the distribution of wealth in society would be further reinforced through the mechanism of price rationing. The rich would be able to afford to give such a drug to their children, the poor would not. Taken to its most extreme, society could polarize between the wealthy and intelligent, with that intelligence artificially boosted by drugs, and the indigent and unintelligent who have no hope of escaping from their condition because their economic circumstances do not permit them to purchase the way out – the drug. The correlation between wealth and IQ is not of course 100 per cent in present-day society! Yet such a polarization is only a matter of degree, and continues inequities already permitted by society, some of which relate to the enhancement of intelligence or at least of intellectual performance by environmental means. Wealthy people can buy what they believe to be a better education and a better future for their children than is available through the state school system. Nor need such discrepancies be as simple and clear-cut as the ability to buy specific goods on the free market: on average the children of the middle classes receive better nutrition and better health care than those in the homes of manual labourers and this is reflected in their scholastic achievement. Society already tolerates such disparities in access to education, direct health care, and general health, fitness and enjoyment of life, so why should the availability of an IQ-enhancing drug be viewed any differently?

*

146

This discussion has led away from the privacy of the geneticist's consulting room into the bustle and confusion of traders hawking their wares in the market-place. But this is how it should be, because treatments for genetic disease are already marketable commodities. There are conditions which are so common and widely accepted that they are not often seen as genetic defects and for which the market solution is considered unexceptionable. I personally have inherited one such mildly disabling genetic defect – myopia or shortsighted-ness. In the industrialized countries, an international industry has grown up to cope with such deficiencies. My optician may not like to be so characterized, but he makes his living by exploiting my genetic deficiency and those of many others like me. My genetic infirmity cannot be cured, but it can be corrected, and the glasses that I wear every waking moment allow me to live as close to 'normal' a life as possible. The fashion industry contributes here in the design of the frames, the size of the lenses themselves, or in the vogue for contact lenses. There are still some things I cannot do while wearing glasses – play rugby, for example – but I do not regard that as any great loss. This may seem a trivial condition, hardly even a genetic 'defect' when set beside such life-threatening diseases as CF or sickle-cell anae-mia, but it illustrates that remedying genetic defects is not solely a medical matter. The manufacture of the lenses and of the frames of my glasses is an international mass-production industry.

Too much discussion of genetic disease takes place almost in a vacuum, as if the only players were the physician and his patient. This is a poor model of reality. In fact there are other spectral presences in the consulting room. One, like Banquo's ghost, is the reckoning that is to come: someone has to pay for the treatment that the physician prescribes. In countries with private health care, that third party will be the private medical insurance company which will have restrictions on whom it is willing to insure and what conditions it will pay to have treated. In the case of countries where health care is provided on a national scale, the third party will be the state itself. The fourth presence in the consulting room will be the provider of the treatment. In the case of most genetic diseases, this will be the international pharmaceutical industry. One thing is common to almost all providers of treatment: the desire to make a profit. And that is where the problems start for many of the genetic diseases which have been discussed in this book: the market system is not perfect and its deficiencies are likely to work against the interests of those suffering from genetic disease.

The first problem is scientific rather than commercial. There are only a few conditions, like human growth hormone deficiency, which it is possible to treat by direct administration of a replacement protein. This is the major difficulty in devising a treatment for sickle-cell anaemia. It has not proved possible to develop a synthetic protein to treat this disease. Max Perutz, the Austrian-born scientist whose decades of painstaking work at the Laboratory of Molecular Biology at Cambridge, England, revealed the three-dimensional shape of the haemoglobin molecule and how it functioned, has been trying for years to devise a substitute for the natural carrier of oxygen in the blood, but they tend to be broken down in the body before they can be useful. In March 1992, his group reported some success in devising artificial blood, but it remains to be seen whether it will be completely suitable for clinical practice or whether some unanticipated disadvantage might limit its use.

Contrast sickle-cell anaemia with the case of another inherited disease affecting the blood – haemophilia A. Haemophilia affects about 1 in 10,000 boys. The blood lacks protein which helps normal blood to clot. Without this clotting factor – factor VIII – the boys can bleed to death from a simple cut and suffer internal damage from extensive internal bleeding resulting from everyday bruises and knocks. Factor VIII is comparatively easy to obtain from normal blood and can be given effectively by injection. It is not an ideal solution: haemophiliacs require frequent infusions of factor VIII throughout their lives, and in recent years there have been tragic instances of haemophiliacs contracting AIDS through contamination of the blood from which factor VIII is derived.

There seems to be little prospect of developing a simple pharmaceutical to remedy sickle-cell anaemia or the thalassaemias. The best current treatment remains a bone marrow transplant, if a compatible donor can be found. There are several other conditions which are amenable to this approach but only about one third of patients have a suitable donor, and results are of mixed success. The first bone marrow transplant was carried out in 1968 on a boy with Severe Combined Immune Deficiency. SCID is particularly suitable for treatment by transplantation, because the recipient has no functional immune system to reject the donor's bone marrow cells. But bone marrow transplants from compatible donors are still not problem free. Even if there is no problem with the host's immune system rejecting the new cells, there is the possibility of immune system cells in the donated bone marrow recognizing their new body as 'foreign' and attacking it. Graft versus host disease represents a severe constraint upon the procedure. However, it is uncommon if transplantation is carried out early in life, so the best hope of correcting genetic diseases amenable to treatment in this way is to carry out the transplants as soon as possible after birth.

But for the majority of patients, such options are not available. What remains is palliative treatment, aimed not at a full cure but at alleviating the symptoms and allowing the sufferer to live as near normal a life as possible. In 1985, a study carried out by Johns Hopkins and McGill universities found that such symptomatic treatment was successful in only about 12 per cent of genetic diseases, moderately effective in about 40 per cent, but completely useless in around 48 per cent. One reason is the sheer intractability of most genetic diseases, in scientific and clinical terms. But another factor is that they are in general rare, sometimes very rare, and thus do not represent an attractive market for free-enterprise companies to invest in. Consider sickle-cell anaemia: although it ranks as the second most common genetic disorder in the world after the thalassaemias, to the medical practices and drug companies of the industrialized countries it is a comparatively rare disease. An incidence of 1,500 new cases in the USA each year is just not enough to make it commercially interesting. In general, drug companies are not interested in developing new drugs unless the market is large enough for the product to be profitable.

Following the thalidomide disaster in the late 1960s and early 1970s, most developed countries have instituted very strict controls over the introduction of new drugs: they have to be exhaustively (and therefore very expensively) tested to show that they do not have toxic side-effects before they can be released for prescription. Thus pharmaceutical companies have not only to invest in the research and development which would lead to a candidate drug, but also to pay for the testing that the drug has to undergo before it can be marketed. These are heavy financial penalties, and the company must look for a good profit to recoup its investments. According to Dr George

Brewer from the University of Michigan Medical School, 'Most genetic diseases are sufficiently uncommon so that they do not fall into this category.' In 1979, Brewer reviewed the halting progress of development of treatments for sickle-cell anaemia in the journal *Perspectives in Biology and Medicine*. He points out that there is little interest in developing therapeutic agents for sickle-cell anaemia, because although there are at least 30,000 sufferers in the USA, sickle-cell anaemia does not rank as common enough to be of commercial interest to the pharmaceutical companies there. This leaves the task of drug development to the physician-investigator, Brewer continues, but most such investigators are inexperienced and unskilled at drug development.

A similar situation affects those with thalassaemia. For those afflicted with severe forms of the disease and for whom no compatible bone marrow donor is available, the standard treatment is regular blood transfusions, every three to six weeks. Such treatment can keep severely affected children in comparatively good health for ten or fifteen years, whereas without treatment they would die within three years. But the transfusions lead to a build-up of iron in the body and this excess itself causes damage to the heart, liver and other organs. Many children with the severest forms of thalassaemia die as a result of iron overload before they reach the age of twenty, even though a drug to remove excess iron, desferrioxamine, has been available since the mid-1970s. But desferrioxamine has to be administered by a daily eight-to-ten-hour subcutaneous infusion with the aid of a pump. The treatment is expensive, costing about £4,000 per patient per year, and many patients risk their lives by rebelling against their daily attachment to the pump. Thalassaemia is predominantly a disease of the developing countries where few can afford such expensive treatment, so there is a great need for an iron-removal drug which can be taken orally and which therefore should be cheaper. But only in 1991 did the *British Medical Journal* carry a report on such a compound, which had been developed at a London hospital and which had already undergone some five years of clinical trials. Even so, its future is uncertain: the drug will not become widely available until 1997 and it is not clear if any pharmaceutical company will fully commit itself to it.

In the United States, the government has moved to encourage pharmaceutical companies to invest in developing drugs for less common diseases. The so-called Orphan Drugs Act allows greater protection than normal to the developer of a drug and thus a greater chance for the company to recoup its investment. Such legislation has proved controversial, but it may be the only way in which treatments for many genetic diseases will become available.

The commercialization of the products of the Human Genome Project will be at least as important and throw up at least as many moral and public policy problems as any other aspect of the project. While the geneticists may be seeking all 100,000 human genes with equal eagerness, some genes will be more important commercially than others. Unless action is taken in advance, we could land ourselves in the paradoxical situation that we have acute social problems because of a plentiful supply of a 'cosmetic' protein resulting from one gene (the putative IQ enhancer, for example) while human suffering continues because no one is interested in producing drugs to treat some inherited disease.

The foregoing discussion examined the consequences of isolating human genes and inserting them into bacteria or other micro-organisms to act as protein production factories. The next logical step is to examine what happens when these techniques of genetic engineering are applied to human beings. In this case too, the future seems to be here already. At the beginning of Chapter 2, I recounted the first successful case of human gene therapy where, in 1990, Michael Blaese and his colleagues at the US National Institutes of Health transplanted a correct gene into a girl whose immune system lacked a vital protein – adenosine deaminase. With the miserable track record of conventional therapy for genetic disorders, it is perhaps not surprising that many people are looking to gene transplants as the hope for the future.

There are potentially two types of gene therapy. In one, 'new' genes are inserted only into an individual's somatic cells – these are any cells of the body, such as those of blood or muscle, except for the germ cells, which consist of the eggs or sperm and the cells which give rise to them. In such a case of 'somatic' gene therapy, the effect of the transplant dies with the individual and cannot be passed on to their children. The other type of therapy occurs when new genes are inserted into a fertilized egg so that the individual who subsequently develops carries the transplanted genes in all their cells – eggs or sperm included. Someone who has been the subject of such 'germ-line'

151

gene therapy may thus pass the new genes on to his or her own children.

Used simply to cure disease, somatic gene therapy raises relatively few problems from a moral and public policy point of view: anyone who accepts organ transplantation would find little moral difference between that and the transplantation of a gene. This relatively uncontentious strategy will therefore be discussed first. Moral problems arise when gene therapy is applied not to the treatment of a recognizable disease, but to the enhancement of existing traits; not to rectifying Nature's mistakes, but to improving upon Nature. Germ-line gene therapy raises all manner of ethical difficulties even if the aim is to treat disease, and it will be discussed later.

Although somatic gene therapy may be relatively unproblematic from a moral point of view, it is fraught with technical difficulty and the ethical problems it poses are intimately bound up in these technical matters. The first point is that, in the short term, the promise of gene therapy may have been oversold. The case of ADA deficiency will serve as illustration. The transplant carried out at Bethesda does not represent a permanent cure but resembles the situation in haemophilia. Repeated gene transplants will be required, just as haemophiliacs need recurrent injections of factor VIII. The problem is that the white blood cells into which the correct ADA gene is transplanted do not last for ever in the body. After fulfilling their normal function they are broken down and removed from circulation. New white blood cells are continuously produced from 'stem cells' in the bone marrow. For a gene transplant to be really effective, the genes ought to be inserted into the stem cells and then there would be no need for subsequent transplants. But stem cells are rare and difficult to identify in bone marrow. However, in April 1992, Italian doctors carried out an ADA gene transplant on a boy in which they claim to have found a way of selectively enhancing the transfer of the gene into stem cells. If they are successful, then this will open the way to a permanent treatment for the condition.

Another problem lies with inserting the gene into a suitable site in the cell's DNA. At present it is impossible to direct the insertion of the gene in any particular place in the DNA. If the gene is not in an appropriate place, it may not be properly regulated – it may make too much or too little of its protein. One reason that ADA deficiency was chosen as the subject of the first gene transplant is that the exact quantity of ADA seems actually not to matter too much – as long as some ADA is being made, the immune system will develop properly – so the precise regulation of the gene was not so important in this case. There is a further danger that the inserted gene might interfere with the operation of other genes, the greatest worry being that it could trigger latent cancer-causing genes into action. Clearly, in the case of ADA deficiency, such considerations pale into insignificance against the prospect of imminent death. The fact that the genetically engineered cells would be removed from the girl's system within a matter of months is a point in favour in this case: even if a cell's genetic machinery were disrupted, it would probably not be around long enough to cause problems. On the other hand, if a method of identifying stem cells and carrying out gene transplants into these cells is found, the dangers of triggering oncogenes into life could be that much more real.

Sickle-cell anaemia and thalassaemia are, on the face of it, prime candidates for gene therapy. But the problem of controlling gene expression currently represents the main obstacle. Haemoglobin is produced as a result of the action of two genes: one coding for the α chains within the molecule and the other coding for the β chains. The rate at which the α chains are produced has to match exactly the rate at which the β chains are produced. If gene therapy is to work for, say, β-thalassaemia, researchers will have to insert a new β-globin gene and ensure that it is expressed at exactly the same rate as the α-globin gene. Were production rates not to balance, then the effect of gene therapy might be to convert β- into α-thalassaemia. In treating all defects involving haemoglobin, the genes must be inserted into bone marrow stem cells, for the mature red blood cells themselves do not make haemoglobin at all and only last about 120 days in the circulation. The genes must be expressed only in those cells which are going to produce red blood cells – there would almost certainly be difficulties if, as a result of gene therapy, the patient's white blood cells started to produce parts of a haemoglobin molecule. Technical difficulties of this nature contributed to the failure of that premature attempt to treat thalassaemia by gene therapy in June 1980. Professor Martin Cline, from the University of California at Los Angeles, operated on two patients, both desperately ill with β-thalassaemia: a twenty-one-year-old woman at the Hadassah Hospital in Jerusalem;

and a sixteen-year-old at the Poly Clinic in Naples. The procedure did the patients no harm, but neither did it do them any good. The basic science was just not there: the mechanisms which control the expression of the globin genes are still not fully understood a decade later, although there has been progress.

Other genetic diseases present similar problems, although for some the would-be genetic surgeon has the additional difficulty of working out how to get the genes into the target cells. The bone marrow is relatively easy to work with: if stem cells can be identified, it is not difficult in principle to remove a sample of bone marrow, genetically engineer the stem cells, and replace them in the body by transfusion in exactly the same way as a normal bone marrow transplant. But it is not possible to treat someone with CF by removing the cells that line their lungs, genetically engineering them in the laboratory and then replacing them. Some researchers are working on the idea of using viruses – similar to those that cause the common cold – which naturally infect the lungs and respiratory tract and genetically engineering them to carry an intact CF gene into the cells that line the lung. Again, this would not be a permanent cure, for there is a high turnover rate of these cells too. Moreover, the mucus that clogs the lungs might obstruct the passage of the virus into the affected cells and, most seriously, this treatment would do nothing to remedy the digestive problems which also plague CF sufferers. A very different approach is being considered for boys with Duchenne muscular dystrophy (DMD) – direct injection of DNA into the muscles. For reasons that are not yet clear, muscle cells are particularly good at absorbing DNA and there is some evidence that they will take up the dystrophin gene and start producing the protein.

Several points follow from these technical problems. The first is that, until the difficulty of identifying and isolating stem cells is solved, gene transplants which involve bone marrow cells will have to be done at regular and frequent intervals throughout the patient's lifetime. The second consequence is that in each transplant, the new cells will have to be carefully screened and monitored before being grafted back into the patient to check that they are working properly and will not be liable to disrupt other subtle mechanisms. Thus there is no prospect of being able to standardize this method of treating genetic disease. It will have to be done on a case by case basis. The cost of gene therapy will therefore remain high although it will doubtless fall as procedures become more routine.

An economic question mark thus hangs over the whole business of gene therapy: will those who pay the bills be willing to pay for such heroic medical treatment? In a country such as Britain, will the already overstretched resources of the National Health Service be able to meet demands for gene therapy in the future? Or will such treatments be regarded as yet another piece of glamorous high-technology medicine diverting precious funds away from mundane but more common ailments which the health service must also treat? In the United States, it may be that insurers will decide not to pay for such treatments; they may instead tell their clients that they are willing to pay for prenatal screening of foetuses that are at risk of thalassaemia or sickle-cell disease – but only on condition that any affected foetus is aborted.

In most developing countries, already rudimentary health care systems simply could not stretch to cover gene therapy for all children born with these diseases. Although such countries will no doubt have personnel skilled enough to carry out the prenatal diagnosis, they may still lack the people who could carry out the therapy. The dichotomy, which is not unique to genetic medicine, would once again come into force: the children of the rich might get such treatments whereas the poor would have to make do with prenatal diagnosis and selective termination of affected pregnancies, or with nothing at all.

Many governments have shown themselves indifferent to the deaths of thousands of their adult citizens by persecuting and killing their political opponents. It is unlikely that such governments would have much concern about coercing their citizens into programmes for the premature termination of the as-yet unborn, in order to save money on the national health budget. Perhaps the only effective means by which such pressures could be resisted might come, in some countries, from religious or social taboos against abortion.

A predilection for interfering with the personal decisions of its citizens about having children need not be confined to despotism. The experience of family planning programmes in India in the 1970s demonstrated how great can be the intervention of government in the most personal and private areas of its citizens' lives. In the world's largest democracy, sterilization was forced on many people by the authorities and the rights of the individual were over-ruled in what was believed to be the collective interest. It is possible to draw some comfort from the fact that such coercion ultimately failed, but not before lasting suffering had been inflicted on many people.

Many of the ethical questions posed by gene therapy have been examined publicly. In the United States, in June 1980 (the same month in which Professor Cline carried out his unsuccessful attempt at somatic human gene therapy), the three main religious groups in the USA – the US Catholic Conference, the Synagogue Council of America, and the National Council of Churches – wrote to then President Jimmy Carter, to express concern about the rapid advances in genetic engineering. In response, the President's Commission for the Study of Ethical Problems in Medicine and Biomedical and Behavioral Research undertook a study of the issue which resulted in the publication in 1982 of a comprehensive report: *Splicing Life: the Social and Ethical Issues of Genetic Engineering with Human Beings*. The commission concluded that although gene splicing was a revolutionary scientific technique 'it does not necessarily follow that all its applications or objectives represent a radical departure from the past'. In particular, it could find little of contention in somatic-cell gene therapy which 'would resemble standard medical therapies in that they all involve changes limited to the cells of the person being treated. An analogy is organ transplantation, which also involves the incorporation into an individual of cells containing DNA of "foreign origin".' It is worth noting one aspect of gene therapy which, in 1982, the President's Commission thought distinguished it from other treatments but which from the standpoint of the early 1990s does not seem so likely. The commission thought that 'gene therapy involves an inherent and probably permanent change in the body rather than requiring repeated applications of an outside force or substance.' That aspect of gene therapy has not yet been realized.

The 1982 report by the President's Commission was followed by a background paper from the Congressional Office of Technology Assessment which concluded that 'Civic, religious, scientific and medical groups have all accepted, in principle, the appropriateness of gene therapy of somatic cells in humans for specific genetic diseases. Somatic-cell gene therapy is seen as an extension of present methods of therapy that might be preferable to other technologies.' Following these studies of the issue, in September 1986, the NIH drew up a list of 'points to consider' for anyone who wished to approach the NIH for financial support to carry out somatic gene therapy. Again there was little in the way of moral objection to somatic cell therapy for the purposes of alleviating disease. Separate national investigations into the issue in Sweden, the Federal Republic of Germany (as it then was) and France came to very similar conclusions, as did the much later British *Report of the Committee on the Ethics of Gene Therapy* which was published in 1992.

The first point, which the British, American and other investigations are agreed upon, is that gene therapy is still very much an experimental procedure. The professional and ethical norms to be applied to gene therapy are therefore those applied to research conducted on human subjects, not those that apply to routine medical practice. The professional constraints on medical research with human subjects is a long-established branch of medical ethics and will not be considered in any detail here. However, the British study identified at least one point at which the specifically genetic nature of the procedures makes a difference. Any doctor who carries out an operation on a human subject has a duty to follow up and monitor what happens subsequently, to assess and if possible mitigate any adverse consequences of the operation. In a case of gene therapy, the British committee recommended, this duty of follow-up care will not end with the death of the patient – to ensure that somatic gene therapy has not inadvertently affected the patient's offspring, monitoring must continue for at least the next generation. This is a much tighter condition than seems to be required in the USA under the NIH's 'points to consider' which talks only of three to five years as the period of 'long-term follow-up'.

The British committee noted that the potential uses of gene therapy might range more widely than the correction of single-gene disorders. 'For example, it is being investigated as a possible new approach to the management of a wide spectrum of diseases ranging from infections such as AIDS to cancer, and it is being studied as a means of strengthening the body's immune response to viral infections.' Although such 'wider applications may soon call for attention', the committee concluded that 'in the current state of knowledge it would not be acceptable to use gene modification to attempt to change human traits not associated with disease'. In the US, the NIH guidelines concede that 'while research in molecular biology could lead to the development of techniques for the use of genetic means to enhance human capabilities rather than to correct defects in patients, [the NIH] does not believe that these effects will follow immediately or inevitably from experiments with somatic-cell gene therapy.'

All the national studies of the ethics of gene therapy have rejected, at least for the time being, the idea of germ-line therapy. In France and Germany, the recommendation was for outright prohibition. In Sweden, the conclusion was that any proposals for germ-line therapy would have to come under severe ethical examination. In the USA, the NIH 'points to consider' stressed that the NIH 'will not at present entertain proposals for germ line alterations but will consider for approval protocols involving somatic cell gene therapy'. The British committee ruled out germ-line therapy on the grounds that 'there is at present insufficient knowledge to evaluate the risks to future generations'.

The prohibition on germ-line therapy applies only when humans are the subjects. The technique is already commonplace in the laboratory and 'transgenic' sheep and cattle can be seen grazing peacefully at farms attached to research institutes in many countries. One straightforward way of introducing 'foreign genes' into laboratory mice is by direct injection into the egg. A very fine needle containing a solution of the DNA to be added is inserted into a newly fertilized egg. After this microinjection, the embryo can be transferred to the uterus of a surrogate mother. However, the procedure can damage the egg, so only a small proportion of the injected embryos develop. Moreover, the injection does not always 'take': in only a fraction of the embryos does the injected gene integrate itself properly into the DNA of the mouse. A further limitation is that sometimes integration only happens after the mouse's genome has replicated several times, so the additional DNA is present in only some of the cells of the adult mouse. It should be clear that many of the difficulties mentioned earlier with respect to somatic-cell gene therapy – particularly with respect to controlling the site of integration – apply also to germ-line therapy.

The astonishing thing is not that there are so many difficulties and uncertainties resulting from microinjection of foreign DNA into fertilized mouse eggs, but that the procedure works at all. None the less, it is clear that while a high failure rate may be acceptable when dealing with laboratory animals, no one would be prepared to tolerate any failure were germ-line therapy to be applied to a human egg. These technical considerations by themselves rule the procedure out for many years to come.

But there is a further technical point relevant not just to the individual who develops from an egg which has been manipulated in such a way, but to that individual's own progeny. No gene therapist at present can cut out a defective gene and replace it with an intact version; they can only add an extra copy of the 'correct' gene to someone's DNA. So if therapy were conducted on a fertilized human egg, then the individual who developed from that egg would still carry the defective gene in their DNA as well as the extra corrected gene. They would thus be at risk of passing on the defective gene to their children, even though the purpose of germ-line therapy is to influence the germ line and prevent further transmission of a defective gene. It is difficult to see what advantage germ-line gene therapy holds over somatic cell therapy.

Yet the issue of germ-line therapy is hotly debated and, perhaps surprisingly, prominent among those who appear most willing to endorse its future use are eminent moral philosophers. Writing in the Spring 1992 issue of *Science and Public Affairs*, the British philosopher Baroness Mary Warnock contemplates the acceptability of germ-line therapy not to cure disease but as a means of enhancing normal characteristics. She writes:

If it became possible, as indeed it might, to eradicate for ever immune deficiency disease, in particular AIDS, through germline therapy, the present or immediate advantage might seem sufficiently great to outweigh the argument from ignorance (however keenly felt). I would not like to rule out for ever the legitimacy of germline genetic manipulation at the embryonic stage.

In passing, it is worth remarking that there is no clear reason why defence against AIDS should be such a compelling reason to contemplate germ-line therapy. One might consider that in terms of global suffering, it will be a long time before AIDS comes close to that inflicted by sickle-cell anaemia and thalassaemia. Were technology to overcome the obstacles outlined earlier in this chapter and make gene therapy possible to remedy these diseases, one might consider that they would be as morally deserving of attention as an infectious disease which happens to have generated panic headlines in the newspapers of the industrialized countries.

Discussions of germ-line therapy inevitably encroach on to the territory of the deeply hypothetical, but it might be worth considering germ-line intervention to eliminate sickle-cell disease and thalass-aemia a little bit further. This could have two components; one would be to genetically engineer back to health an embryo which had inherited two copies of the deficient globin gene and which would otherwise be born with the disease. If techniques for cutting out and replacing defective genes had been fully developed, the person who developed from this embryo would never suffer from the disease and would be free of the risk of passing on a single copy of the defective gene to his or her children. It is difficult to find any moral objections to this procedure which is clearly aimed at curing disease and not at enhancing normal traits.

Moral problems do, however, arise when one comes to consider the second level of germ-line intervention, which would be to 'treat' an embryo carrying a single copy of the gene. The aim of germ-line therapy here would be to eliminate the risk of there ever being people who might pass on the gene to subsequent generations. It is worth recalling that people who carry a single copy of the sickle-cell gene are perfectly fit and healthy, have an increased resistance to malaria, and are at risk of having children suffering from the disease only if they marry another carrier of a single copy of the gene, and even then the risk of an affected pregnancy is one in four. Germ-line therapy to eliminate the gene is thus interfering with what might be considered a reasonably normal trait. If the practice became widespread, it would represent a major alteration to the gene pool of the human race as a whole.

Two problems arise when one considers such radical intervention in the human genetic structure. The first harks back to the mess that the United States made of screening the population for sickle-cell disease. Carriers of a single copy of the gene were wrongly labelled as being sick and were discriminated against in health insurance and employment. If a programme began to 'treat' the condition by genetic surgery, it is difficult to see how, even with the best preparations and public information campaigns, it would be possible to avoid stigmatization of anyone who had not, as an embryo, been operated on in this way. Such a programme might begin as a voluntary one, but pressure from vested commercial interests (health insurers perhaps, or the state) could easily transform it into one where overt or covert compulsion to undergo the treatment inevitably ensued. The twist to this argument is that of course no one could give consent to being operated upon when they were in the form of a fertilized egg. The decision would have to be taken and the consent given on behalf of the future person by the couple who had conceived the embryo.

A second argument against such a programme depends upon the appeal to ignorance which Baroness Warnock cited. Proposals to eliminate a specific gene from the gene pool altogether revive the old, largely discredited notion that there are such things as 'good' genes and 'bad' genes. The truth is that we know little about the natural history of human genes. In the case of thalassaemia and sickle-cell disease, it is clear that these variants arose as a direct evolutionary response to the threat of malaria. It is also clear that they do not represent a very efficient defence against the disease and that changing the environment, by eradicating the mosquitoes which transmit the disease, is a better option. Sadly, this does not appear to be happening at present: there is a resurgence of malaria around the world. Mosquito eradication programmes are breaking down in Africa in the aftermath of civil war and the collapse of ordered government in many African states, and the malarial parasite itself is becoming resistant to some antimalarial drugs. But even if malaria were controlled, it would remain the case that not enough is known about the genesis of the sickle-cell and thalassaemia variants for it to be safe to contemplate eliminating them from humanity's gene pool. This argument applies *a fortiori* since, if techniques have developed sufficiently to conduct germ-line therapy, it will certainly be possible to treat by somatic gene therapy those who are actually suffering from the disease.

Another British moral philosopher has suggested that germ-line gene therapy to enhance human traits is something devoutly to be wished for. In his book *Wonderwoman and Superman*, John Harris, professor of applied philosophy at the University of Manchester, imagines a new breed of humans created in the laboratory with their genes reinforced so that they are immune to diseases such as AIDS, malaria and hepatitis B. In addition, he confers on his imaginary creations genes which give them better resistance than average to cancer, and which allow them to live longer but without senescence. He writes: 'For my own part, I welcome the possibility of a new breed of persons with life chances not available to us now.' Yet while

Professor Harris supports this conclusion with extensive moral argument, he fails to address the most important questions: who will provide such enhancement services, and who will control and regulate the providers? These are central to all discussion of the applications of the new genetics and more so to the prospects of germ-line therapy.

Money, medicine and power are all at issue. The first, and perhaps least important, point is that germ-line therapy will be inescapably expensive, so that questions arise of who would pay for the operations to be carried out and of whether this is a wise use of scarce resources. These questions have arisen at various points in earlier discussions of applications of the new knowledge which the Human Genome Project will bring about, and the answers are likely to be the same: that the rich will be able to buy preferential treatment on the free market while comparable treatment for the poor will be rationed, if it is available at all; that only careful accounting of the benefits and costs would be able to establish if such procedures were a wise use of scarce national resources; and that the existing economic institutions in a country may swing the balance in a direction which is not optimal for the nation as a whole.

The second point is that germ-line therapy would require a corpus of highly trained genetic surgeons and that, inevitably, such people will be in short supply. Yet they will have technical control over the future genetic composition of the human race. The future of humanity's genes has hitherto been left to the gloriously messy and random business of men and women choosing their sexual partners (not always the same thing as choosing their marriage partners). Will people be willing to surrender control of this aspect of their lives to a small élite technocracy? Baroness Warnock is alive to the problem. She writes: 'The real reason for the general assumption throughout Europe that germline intervention will be prohibited is the fear of the power of doctors. We do not like the idea that someone unknown, but certainly not we ourselves, will be able to exercise power over not only us but our children by choosing, albeit a bit at random, how we should be.' She then goes on to make the final point in this matter, that 'we all fear, and not without reason, that one day such power might be exercised, not by benevolent doctors, but by political tyrants who would use us for their own ends.'

The history of the social impact of human genetics leads inevitably to the conclusion that Warnock has both overstated the benevolence of doctors and neglected the fact that democracies, not just tyrants, have a poor record in dealing humanely with human genetics. Before the Second World War, the idea that humanity could be improved by meddling with its genes was remarkably popular. The eugenics movement was powerful and influential on both sides of the Atlantic and eminent geneticists were at the forefront. Its origins can be traced back at least to 1865 and Francis Galton, professor at University College, London. The stated aim was 'to check the birth rate of the unfit and improve the race by furthering the productivity of the fit by the early marriages of the best stock'. The technologies of direct intervention in DNA were of course not available, so the mechanisms by which the genetic improvement of Man (never Woman) was to be brought about were those of the selective breeder in the farmyard and racing stable.

In Britain, a society dominated by divisions of social class, eugenic ideas roused anxieties among the upper classes that the lower orders with their high birth rate were outbreeding their betters and that the general level of intelligence in the population would fall. (Given the historical record of many members of the British aristocracy, this argument may seem something of a *non sequitur*.) Even as late as 1958, so great a scientist as Sir Peter Medawar devoted much of his Reith Lectures on BBC radio to discussing the possibility that the average level of intelligence might decline because (although this was not stated quite so crudely) the lower classes were no longer dying sufficiently quickly, thanks to the post-war Welfare State, to balance their fertility rate.

In the 1990s, the desperately troubled state of the English education system, while the consequence of many other factors, certainly derives in part from the assiduous efforts of educational psychologists in the 1930s and 1940s who believed firmly in the heritability of IQ and in a correlation between intelligence and social class. Such beliefs contributed to the idea that ability was determined sufficiently early in life for children to be tested at the age of eleven to see whether they would benefit from an academic style of education or whether they should be consigned to the refuse heap and destined for vocational training in manual skills. The divisions thus opened up have not been eradicated from the educational system, nor even from thinking about education.

In the United States, on the other hand, the eugenic movement got bound up in issues of race and in controlling or stopping the immigration to the USA of what was perceived to be inferior stock. In America too, there were concerns about intelligence levels of the population, but the US took its concerns about 'feeble-mindedness' to much greater extremes than did the UK: between 1905 and 1973, nearly 100,000 'feeble-minded' women were compulsorily sterilized to prevent further defective children. Even this issue was confused with race: in the 1920s, one anthropologist prominent in the US eugenic movement, Charles Davenport, tried to relate feeble-mindedness to Negro features by analogy with the way that Down's syndrome had originally been described as mongolism because of a superficial facial similarity between those with the syndrome and people from Mongolia.

Eugenics had two features in particular: one was that concern about future people took precedence over the interests, autonomy and rights of the living; the second was the concept that the interests of the race were more important than those of the living individual and indeed of potential individual unborn children. Thus in the United States the power of the state was employed to coerce people, on the basis of their presumed genetic make-up, into behaving in a way that they would not personally have chosen. A better understanding of human and general genetics gradually undermined the eugenicist position and the appalling episode of the Second World War, when Nazi ideologues put a racial hygiene programme into action, finally destroyed its credibility. But it is worth recalling that although the worst excesses of eugenics were indeed committed under a totalitarian tyranny, the record of two of the most firmly established democracies is not one to be proud of. Warnock's 'benevolent doctors' proved themselves willing agents in the sterilization programme in the USA and, in both countries, eminent scientists exercised an influence that was anything but benevolent in the social application of their now-discredited theories.

Some of the ideas of the eugenics movement persist in garbled form and condition popular opinion about human genetics today, even though there is not a shred of scientific evidence to support the eugenicist position. History is forever repeating itself – but never in exactly the same way, so this discussion is not to suggest that minorities will again be tyrannized by being forcibly sterilized. It is to

164

suggest that a new technology does not come into a society which is completely free to decide about that technology's use: choices are constrained by pre-existing institutions – political, social, religious and economic – and, in a democracy, by popular expectations and beliefs, even if these are at variance with how the community of scientists or moral philosophers sees the matter. Yet the choices which a society makes today in relation to germ-line gene therapy will have an effect long into the future, possibly long after the society which has made the choice has itself ceased to exist.

It is perhaps fortunate therefore that many technical obstacles remain before germ-line gene therapy is a reasonable prospect. And it is certainly not clear that germ-line therapy would actually be necessary to achieve the ends which Baroness Warnock and Professor Harris regard as desirable. If germ-line therapy becomes technically attainable, somatic gene therapy will certainly also have been perfected. Just as children today are immunized against the major infectious diseases, one could imagine children in the future undergoing a somatic gene therapy session to boost many of the characteristics which Warnock and Harris think desirable. It may be possible to achieve immunity to AIDS and to encourage longevity and resistance to cancer simply by reprogramming the DNA in stem cells of the bone marrow, for example. What would not be possible would be structural alterations, such as changing eye colour: such choices would probably require germ-line therapy. If such boosting of normal traits is ever considered desirable, it might be wiser to proceed with somatic gene therapy and study its effects, not only on the children, but on society itself, before ever making a commitment to the irrevocable step of germ-line therapy.

8

The Moral Consequences of
Molecular Biology

POPULAR accounts of science, and the memoirs of great scientists reflecting on their careers in the glow of retirement, tend to emphasize the high points, from which broad vistas of intellectual territory are laid out for the lay reader with the tantalizing prospect of further riches beyond the distant horizon. Science is portrayed in this way almost as the post-medieval equivalent of a great Gothic cathedral, with the scientist surveying the landscape and the heavens from the loftiest spire, except that in our secular age it is a temple of ideas rather than of stone and glass and it is erected to the god of knowledge of the natural world.

In reality, the routine business of science is just that: routine, containing a high proportion of hard work, tedium, and even boredom. The secular cathedral is never quite complete and, for most of the time, apprentices must approach it not through the ordered calm of a cloister but through the noise and confusion of the builder's yard. As with building real cathedrals, there is a great deal of routine heavy work for the labourers to perform in fulfilment of some remote architect's designs. Many a postgraduate with newly minted PhD can testify to the feeling of having laboured long and hard and having at best added but one small brick to the edifice of knowledge – a brick, moreover, almost indistinguishable from its fellows. Even those who pursue science as a full-time profession must work to an overall design set out by others. At best, perhaps, one can have the joy of shaping a particularly intricate and pleasing carving over a window, or of putting the keystone into place to complete one of the arches.

Every now and again, however, someone comes along and realizes that the structure has become so greatly embellished that the foundations will no longer support its weight and the whole building has to be pulled down and started again to a completely new design. These moments are comparatively rare, but they naturally attract more attention than the routine work: the destruction of a scientific theory which has stood the test of centuries readily catches the imagination of a public previously indifferent to the labour that had gone into the original theory's construction.

This characterization of science as an activity in which periods of 'normal' science are interrupted by the occasional revolution was first formulated by the American philosopher of science, Thomas Kuhn, in 1962. Although its philosophical status is still controversial, as a description of how science is actually done, Kuhn's picture has not been bettered. In normal periods, according to Kuhn, scientists work within an accepted 'paradigm'. They have a common understanding of what questions are important and of how to make sense of the results of experiments in terms of their common understanding. The paradigm determines the overall direction of their research.

In normal periods, scientists act as puzzle solvers, studying problems which are generated by and whose solutions expand the paradigm within which they work. He who carves a gargoyle does not seek to redesign the church. In periods of revolution, however, the paradigm no longer suffices; experiments, rather than confirming and extending accepted theories, begin to generate anomalies which cannot be accommodated within the existing framework of belief. Different, incommensurable ways of seeing the world begin to arise, and the once homogeneous community of scientists splits into factions, until gradually a new paradigm asserts itself and with it the world view of science is utterly transformed.

The contrast between the paradigmatic and revolutionary periods highlights the almost schizophrenic character of science: most of the time it is a deeply conservative undertaking which explores the consequences of the accepted paradigm and, bit by painstaking bit, builds up evidence of the truth of existing theories. In this conservative phase, research not only supports the *status quo* in terms of scientific theories, but tends to be socially and politically conservative too. But when a scientific revolution comes, the destruction and rebuilding are frequently not confined solely to the edifice of science itself. The consequences spill over into society and affect society's self-image; even those with neither scientific training nor interest in science find that their view of their own position in the world has been changed. Scientific revolutions can be deeply unsettling for society.

The most famous of all is the revolution spearheaded by Galileo, Kepler and Copernicus in the late sixteenth and early seventeenth century. The anthropocentric view of the universe had held sway for two millennia: the earth was stationary at the centre of the cosmos and the moon, planets, sun and stars orbited the habitation of humanity. The physical disposition of the heavenly bodies echoed the spiritual dispensation by which God had selected His people for special favour and accorded them particular significance in an otherwise hostile or indifferent world. Yet once Isaac Newton had completed the Copernican revolution by showing how elegantly the universal theory of gravitation could describe a sun-centred system, he relegated our Earth to an insignificant role in the planetary scheme of things and destroyed the symmetry between the way the world was ordered in religious terms and the way in which it was organized mechanistically. Newton himself was a deeply religious man (although his theology was far from orthodox) and there is in principle no reason why the physical structure of the world should reflect the spiritual. There was no strictly logical reason to consider the new cosmology as conducive to atheism, but the Catholic Church had early recognized the peril it faced from the revolution in science (which followed so closely after Luther's religious schism) and in 1632 prohibited Galileo from promulgating the new ideas. This last despairing attempt to keep humanity swaddled in the comforting superstitions of the middle ages was spectacularly unsuccessful.

Darwin's revolution in the nineteenth century also demoted humanity's place in the scheme of things and again precipitated a dispute between church and science from which religious faith came off the worse. I want to argue that, in some respects, Darwin's revolution is unfinished and that, certainly in terms of its social consequences, the Human Genome Project will come to represent a working-out of the implications of that revolution.

Darwin lacked a theory of inheritance. His theory of evolution required two things: that populations of plants and animals should contain individuals who differed from each other; and that natural selection should act upon these variations, allowing some individuals a greater chance of surviving to maturity and so of reproducing themselves. But although Darwin could give a detailed account of how the mechanism of natural selection worked, he was unable to explain how variations could arise in populations or how these characteristics could be transmitted from one generation to the next. Unknown to Darwin, the Bohemian monk Gregor Mendel was

168

making the first steps to understanding this very topic, the subject we now call genetics, at almost the same time as the publication in 1859 of the *Origin of Species*. Mendel's results, announced in 1865, went unnoticed until they were rediscovered in 1900.

The science of genetics is interesting in its own right, but more than this it is an integral component of our modern-day understanding of the evolution of the living world. The Human Genome Project thus finds its place in a line of scientific development whose antecedents stretch back to Darwin and Mendel. Consequently, it may appear as an example of paradigmatic science. The researchers teasing out the structure of human genes are working according to agreed criteria and common assumptions which were settled in the middle years of this century by the great discoveries that DNA was the molecule of inheritance and that the hereditary information was encoded within its double helix structure. Those discoveries were in turn based on what is now called classical genetics (to distinguish it from the molecular approaches of our own time) whose foundations were laid by Mendel. But within the practice of science, it is clear that the Human Genome Project will change, and indeed is already changing, the nature of genetics research. For the specialists, the Human Genome Project may indeed turn out to be a revolution. Just as today's cosmologists no longer have to grind their own lenses and build their own optical telescopes as Galileo did, so future geneticists will no longer have to go to the labour of stitching a piece of DNA into bacteria in order to find out what function it has. Instead, they will be able to run computer programs to match its genetic instructions with the known sequences contained in databases held in electronic computers. The project has received a mixed reception from the community of geneticists – some scientists actively oppose the idea of earmarking funds for such a narrowly focused approach to genetics research – and one of the reasons may be a fear or suspicion of precisely this sort of change in the nature of the science which the project will bring about.

Although the changes which it will bring about in scientific practice will amount to one of Kuhn's lesser revolutions when compared to developments such as Darwin's theory of evolution, in the realm of social affairs the Human Genome Project has the potential to transform our society and our view of our own place in the scheme of things. There will be many practical consequences – for patients seeking medical care, would-be parents planning a family, prudent investors seeking to save for a pension in their old age, for insurance companies trying to assess the actuarial risk of giving someone life assurance, and for prospective employers assessing the health and capabilities of their workers.

In the end, however, it may be that the Human Genome Project's most far-reaching consequences will be none of these. Its real significance may be that it will mark the culmination of the revolution begun by Darwin and Mendel more than a century ago and that both in scientific terms and in terms of its effect on society, the Human Genome Project may come to fulfil for twentieth-century biology a role equivalent to that of Newton's formulation of the laws of motion and of celestial dynamics for eighteenth-century physics. After Newton, astronomy and mechanics were different sciences to what had gone before: there were new calculations to be done and the old ways of thinking and of doing things were no longer scientifically fruitful. Although Copernicus, Kepler and Galileo initiated the sixteenth century's scientific revolution, its full impact on the rest of society and on humanity's image of its place in the cosmos was not apparent until after Newton. It was Newton's work which laid open the path to an image of the universe as celestial clockwork in which cause inexorably precedes effect according to a rule of natural law, regular and predictable, rather than according to the whims and interventions of a god personally concerned with the products of his creation.

As the Human Genome Project proceeds, it will progressively accumulate new knowledge about the genetic composition of human beings. Even without the project, researchers would be producing such knowledge, but the project will harness their concerted and co-ordinated energies and will so accelerate the accumulation of knowledge as to put strains on the abilities of the rest of us to cope with the flood of information and make sense of the overall picture. Geneticists already have a broad idea of the general thrust of that knowledge. When the details are filled in, and when the information overflows from the laboratories into the clinics of our hospitals, and percolates into the mainstream of general life, then society will have to cope, not just with the new technology but with the consequences of the new knowledge.

Some of the consequences may be quite subtle and unexpected. For example, as individual human genes are identified and analysed, most of them will be found to be similar to genes already known from other animals. It will become inescapably clear just how closely related human beings are to the rest of the animal kingdom. Our common heritage with the rest of the animals has been a commonplace among biologists since Darwin, but it still generates some unease among the lay public. Once the Human Genome Project is complete, that close relationship between humans and the rest will be demonstrated, not just at the level of comparative anatomy of the whole animal (the sort of evidence upon which Darwin and his successors had to rely), but through the comparative anatomy of their and our genes as well. Will such developments cause us to rethink and reassess how we use animals for our own ends and perhaps to conclude that many things which we currently do to animals are no longer morally acceptable? Or will cumulative reminders of our animal ancestry reduce the respect human beings have for each other's humanity?

Whereas one consequence of the Human Genome Project will be to remind us of our similarities to the animals, a second consequence, paradoxically, may be to stress the differences between one human being and his or her neighbour. It is a commonplace that differences of hair or eye colour are genetically determined. But while few people put too much stress on the colour of an individual's eyes, there are other genetic differences, such as skin colour, which have become associated with enormous social and economic differences. Still other human traits, such as intelligence or sexual orientation, have been claimed as being genetically determined (although the evidence is controversial in the extreme) and those who do not conform to the social norm or expectation are severely disadvantaged and discriminated against in many societies. Biologically, the human species benefits from diversity, partly because of what is sometimes called 'hybrid vigour', but more importantly because that diversity represents a valuable pool of genes upon which to draw when, for example, a new disease suddenly strikes. Human society, however, tends to look askance at diversity and those who are different from the norm have tended always to suffer for it. The second consequence of the Human Genome Project will be to point up differences between individual humans at a genetic level. If the knowledge is not handled properly and seen in its proper, biological context, it is possible that the project may generate information which represents potentially fertile soil for new grounds of discrimination.

One further possible consequence of the Human Genome Project stems not from any likely discovery, but from the project's very existence. We may take an increasingly 'atomistic' view of human beings and indeed of life itself. Under the impact of ever more discoveries about human genetics we may come to define ourselves and our lives in genetic terms and neglect the rest. There is a risk that we may all become reductionists – reducing our lives to their supposedly fundamental components – and so may cease to look at them holistically, missing the complexity and richness of life in its entirety.

These then are three possible developments which the Human Genome Project is likely to accelerate and which will be examined in more depth: the similarities between humans and animals will become more apparent; the difference between human and human may be more marked; and the question of the nature of life itself will present itself in a new guise. None of these is a 'practical' matter, where the immediate future can be foreseen in concrete and moderately certain terms. All are in a different category from the development of new tests, new drugs or new surgical procedures, yet taken together they may prove to be as important as the material benefits of the project. To analyse the first of them, the relationship between men and animals, it is necessary to go back to Darwin and Mendel, and to the social impact of those developments.

Publication of the *Origin of Species* in 1859 provoked an immediate clash between Darwinism and the traditional religious doctrine of the Church of England. Until Darwin, it had still been possible for Christians to believe that while the place of the Earth in the universe was unremarkable, nonetheless humanity had been specially created in the image of God at the apex of the living world upon the Earth and so was still a unique, divinely favoured creature. Darwinism, with its idea that humans shared a common ancestor with the apes, destroyed this last hope of finding a central place for humanity in the natural world to reflect the human race's supposed importance in the religious and spiritual world. A distinctive aspect of the theory of evolution, and one which has often been misunderstood, is that it contains no sense of direction or of progress. Darwin himself tried to avoid suggesting that one animal was somehow 'higher' in the biological order of things and another one 'lower': there was only 'fitness', in the sense of being well adapted to the contingent local environment. In this light, it was not even possible to salvage a variant of the religious view by arguing that evolution was but the hitherto unknown mechanism by which God chose to realise His purpose of bringing into existence the human race in His image.

But what is not often realized is that modern evolutionary biology presents problems to traditional morality as well as to religion. The point has been cogently argued by the American moral philosopher James Rachels in his recent book *Created from Animals*. Rachels points out that, if humans are descended from a common ancestor with the apes, this not only destroys the foundations for religious belief in an especial divine favour for humans, but also undermines any special moral value consequent upon being human. In the elegant prose of his *History of Western Philosophy*, Bertrand Russell seems succinctly to have anticipated Rachels' idea:

If men and animals have a common ancestry, and if men developed by such slow stages that there were creatures which we should not know whether to classify as human or not, the question arises: at what stage in evolution did men, or their semi-human ancestors, begin to be all equal? A resolute egalitarian [...] will find himself forced to regard apes as the equals of human beings. And why stop with apes? I do not see how he is to resist an argument in favour of Votes for Oysters. An adherent of evolution should maintain that not only the doctrine of the equality of all men, but also that of the rights of man, must be condemned [...] since it makes too emphatic a distinction between men and other animals.

Russell does not appear to have pursued this line of reasoning or to have examined just how unemphatic is the distinction which can be made between men and other animals. Whereas Darwin and most of his successors based their ideas of common descent on the anatomical similarities and differences between different species, over the past couple of decades it has been possible to read off such relationships at the level of the chemical composition of proteins and the structure of genes themselves. Much of the history of human evolution lies written in human DNA.

Some of the results already thrown up by research into the human genome are quite strange, at least to those who have not been trained in genetics. One is the way in which the genetic make-up of men differs from that of women. Males carry at least one gene which 'switches on' a developing embryo to turn into a male rather than a female. Its discovery and identification by a team in London took years of painstaking research. The human and the analogous mouse gene were discovered in 1990 and the similarity between the two is a small reminder of the common genetic heritage of the two species. In 1991, in an amazing feat of genetic engineering, the researchers demonstrated that it was indeed the genetic switch by putting the mouse gene into an embryo which then developed into a male mouse even though it had been conceived as a female. Whether for mice or men, the gene represents the recipe for a particular protein and the researchers were able to deduce the composition and estimate the functions of this protein. They found that the protein binds to other regions of DNA and thus controls the way in which other genes express themselves. Unexpectedly, they found that it had many similarities to a DNA-binding protein previously identified in yeast cells. The yeast protein is also involved in the 'sexual' reproduction of yeast cells, but that is believed to be a pure coincidence.

The 'switch' for maleness is a recent example, one of the early products of the Human Genome Project, but many more instances are already known of genetic sequences which have been 'conserved' down the years of evolution, usually because the protein product is vital for the functioning of any living cell, whether it be human, horse or yeast. The discovery of these examples predates the Human Genome Project itself and came about mainly by researchers focusing on resemblances and differences between proteins (the gene product) rather than on the analysis of individual genes themselves.

It has been known for decades that the Rhesus blood-group gene is an ancient one which we share with several groups of monkeys. Research on DNA has revealed that humans are even more closely related to our nearest relatives amongst the great apes, chimpanzees, than had previously been thought. By one method of estimation of genetic relatedness, human DNA and the DNA of chimpanzees differ by no more than about 1.6 per cent overall. Genetically, people are closer to chimps than are gorillas; indeed chimps and humans are more closely related than some species of gibbons are to each other.

All this may, to a professional biologist, be stating the obvious. Of course humanity's common ancestry with the apes and the rest of the animal kingdom will be reflected at the level of proteins and of genes. What is new? One answer is that although this may be a commonplace to the professional, it is news and far from trivial news to those without specialist knowledge of modern molecular biology. A second

174

answer is that there is a great deal of difference between the abstract knowledge of a concept – that humanity's common ancestry with the rest of the living world is manifest in human genes – and having that relationship spelled out in full detail by the exact analysis of just how similar each human gene is to its counterpart among animals, plants, yeasts or bacteria. The Human Genome Project will, moreover, explore the full extent of the relationship. A vital part of the whole enterprise is to analyse the genomes of other organisms: bacteria, yeast, the fruit fly, the nematode worm and the mouse will all be examined in detail. In addition, there is considerable commercial interest in teasing out the genes of crop plants: Japan, for example, is particularly interested in the genetics of rice; other countries are looking at maize, wheat, tomatoes and potatoes. As well as laboratory animals, other researchers will be examining the genomes of domestic animals such as cattle and pigs, in the hope of ultimate commercial gain through developing new and more productive breeds. Will there be any genes at all which turn out to be distinctively human and without precedent in the rest of the plant and animal kingdoms?

The details of the genetic relationship do matter, as a final example shows. In 1969, the American geneticist Carl Woese and his colleagues decided to examine similarities within the genetic material of as many organisms as they could in order to construct a universal family tree stretching back to the putative 'progenote' or original ancestor from whom all life on the planet is descended. For technical reasons, the researchers chose to study not DNA but a particular form of the closely related chemical RNA, known as 16S RNA. By 1977, they had their family tree or phylogeny, and according to the standard textbook *Molecular Biology of the Gene*, 'The universal phylogeny based on 16S-type RNA reveals startling conclusions. Human beings are in fact more closely related to corn [maize] than a Gram-negative bacterium (*E. coli*) is to a Gram-positive bacterium (*Bacillus subtilis*) [...] Thus the evolutionary difference separating two different bacteria can be greater than the distance between a sophisticated plant and the most sophisticated animal.' There is no reason to suppose that as the Human Genome Project gathers pace, such conclusions will be overturned; quite the reverse, the project is likely to reveal ever more and ever closer similarities between *Homo sapiens* and the rest of the living world. Increasingly, the unique characteristics of human beings, from which on a traditional view we derive our unique moral value, may come to be seen as the consequence of statistical happenstance: a couple of minor changes in this gene here, a slight mutation in that gene there.

It is by no means obvious that changes in our understanding of the natural world should have any implications for our moral or religious or political beliefs. Logicians have long argued that the way the world is implies nothing about how the world ought to be. The first and most celebrated exponent of this line of reasoning was the Scots philosopher David Hume whose discussion is often abbreviated into the tag: 'is' does not imply 'ought'. The idea was further elaborated by the Cambridge academic G. E. Moore at the beginning of this century who criticized those who underpinned ethical arguments with observations about the world: they were, he said, guilty of 'the naturalistic fallacy'. Yet the issue is not settled and these conclusions have been fiercely criticized by other philosophers who have filled learned journals with articles carrying titles such as 'On how to derive "ought" from "is"'. But whatever the force of logical argument, few of us live our lives by the austere dictates of logic, which tends to play a part in the concerns of real life similar to that which, according to Einstein, mathematics played for physics: 'As far as the laws of mathematics refer to reality, they are not certain; and as far as they are certain, they do not refer to reality.' Increasing knowledge of the closeness of our relationship to the animals may not entail any moral conclusions in a strictly logical sense, but humans tend to be 'reasonable' rather than logical in the way they conduct their lives. And the basic grounds on which it is reasonable to erect a case for the moral superiority of humans over the other animals are being eroded by the accumulation of evidence of our closeness to them.

According to James Rachels, it is now difficult for traditional morality to avoid a charge of 'speciesism'. If claims that humans have a greater moral value than animals were reproduced with 'whites' and 'blacks' in place of 'human' and 'animal', they would today be condemned as racist, and if racist arguments are considered to be bad arguments then, according to this analysis, so must be the logic underpinning traditional morality. Traditional morality asserts that humans and other animals are in different moral categories; that the lives and interests of human beings are of supreme moral importance, while the lives and interests of other animals are relatively unimportant. This doctrine once rested on a religious view of human nature –

THE MORAL CONSEQUENCES OF MOLECULAR BIOLOGY

that man is created in God's image. There are also secular grounds for this view of humanity's moral importance – the idea that man is a uniquely rational being. After all, is it not true that only humans can think, form abstract concepts, and use language? Since moral philosophers depend upon their own cleverness in the use of language when debating the criteria for a creature's moral value, they tend commonly to regard such command of language and abstract thinking as the supreme index of rationality. But Darwin himself argued that if a different test of rationality is applied, in terms of appropriately modifying one's behaviour in response to stimulus or in the face of changes in the environment, then there are clear signs of 'rationality' in all manner of creatures.

Rachels goes on to outline an alternative morality 'without humans being special' in which a wider moral universe is opened up, somewhat as the Copernican revolution opened up a wider physical universe. Humans and nonhumans are, in a sense, moral equals, he argues, and the interests of nonhumans should receive the same consideration as the comparable interests of humans. However, this does not mean that a mosquito should be treated equally to a monkey. 'If we compare the two, we may find that the life of the monkey is far richer and more complex than that of the fly, because the monkey's psychological capabilities are so much greater [. . .] In the light of all this, we may conclude that it is better to swat the fly.' He concludes that it is the richness and complexity of the individual life that is morally significant. He distinguishes the concept of 'being alive' (mere existence) from that of 'having a life' (in which the subject of that life experiences the world, acts within it and, so to speak, constructs a biography for himself or herself). Although biological life is a prerequisite for a biographical life, moral significance attaches only to the biographical life.

The implications of these conclusions are disturbing for traditional ethical beliefs. There is no bar to suicide, if an individual believes that his or her biographical life is valueless, nor to euthanasia; and humans who possess only the capacity for biological life (such as those born with gross mental retardation) may be of less moral value in this scheme of things than a normal chimpanzee or orangutan. Under traditional morality, Rachels writes, 'The biological life of a Tay-Sachs infant, who will never develop into the subject of a biographical life, may be treated with greater respect than the life of an intelligent, sensitive animal such as a chimpanzee.' He concludes that this judgement 'is mistaken'. An even more disturbing corollary is that it might be morally more permissible to carry out scientific experiments involving vivisection on a grossly handicapped human baby than on a standard laboratory animal such as a rhesus monkey. Indeed, Rachels identifies a moral dilemma at the very heart of the case for experimenting on animals: the justification of such experiments is the applicability to humans of the knowledge thus gathered, yet the more closely the experimental animal resembles a human being, the less morally acceptable it is to experiment on it. With its potential for immense practical impact on our society, the Human Genome Project will inevitably propel such questions into the public domain. As the project gathers pace and as the consequences of its researches come more and more to public attention, the moral significance of our common genetic inheritance with the animals will become inescapable.

If one effect of the Human Genome Project may be to blur further the distinction between humans and animals, it is possible that a second consequence may be to accentuate the differences between humans. Race, religion, ethnic origin, cultural heritage and even outward appearance have all been used as the pretext for unfair discrimination among humans in society. The pretexts are in plentiful supply and are unlikely to be supplemented by any results from the Human Genome Project. But the project will enhance and encourage the habit of thinking about people in genetic terms rather than as individual members of society, and there is a risk that existing forms of discrimination may find new and specious justification in the language of biological determinism.

Once again, the seeds for this attitude can be found in developments since Darwin. In the late nineteenth century, some individuals, most notably Herbert Spencer, a figure on the fringes of science and philosophy, enthusiastically picked up Darwin's ideas (Spencer, not Darwin, coined the phrase 'survival of the fittest'), but transferred the theory from a description of processes operating in the natural world to a prescription of how society should be organized. In his seminal lecture *Evolution and Ethics*, delivered in 1893, T. H. Huxley, the most energetic protagonist of Darwin's evolutionary ideas, rejected such a move. 'Law and morals are restraints on the struggle for existence between men in society,' Huxley acknowledged and, moreover, 'the ethical process tends to the suppression of the qualities best fitted for success in that struggle.' But for this very reason, Huxley believed, one should reject the idea of an ethic that was 'applied Natural History' and he characterized the 'cosmic process' of evolution by natural selection as pursuing ends which were not human. He believed instead that 'social progress means a checking of the cosmic process at every step and the substitution for it of another, which may be called the ethical process'. The ethical progress of society, Huxley maintained, 'depends not on imitating the cosmic process but in combating it'.

Despite the vigour and cogency of Huxley's attack on those who sought to create moral values out of simple descriptions of the way the world is, attempts to 'biologicize' ethics continued. Darwin's scientific revolution may have had profound social effects in knocking the last significant prop out from under traditional religious belief, but 'Social Darwinism' was very soon being used to justify the imperialist foreign policies of the European great powers, for was it not part of the natural order that the white man's civilization was superior to the unenlightened superstition under which the darker races lived? And while a racist view of other societies was thus erroneously justified, within northern European society similar ideas were developed to show that the rich and other members of the upper classes were somehow biologically superior to their social and economic inferiors. These ideas were employed within Western society to bolster the existing political and economic order, to justify gross inequities of wealth and poverty, and as an intellectual well-spring for the eugenics movements of the first few decades of this century. This corruption and misapplication of science reached its nadir in the racist ideology of Nazi Germany, although the basis of Nazi ideology is much more complex, drawing heavily as it did on elements of Teutonic mythology as well as the philosophy of Nietzsche, who was contemptuous of Darwin's ideas.

Over the past couple of decades, a veritable industry known as sociobiology has grown up which seeks to interpret the behaviour of human beings in society in terms of basic biological responses driven by the evolutionary imperatives of the struggle for survival and the need to reproduce. Once again, when such analysis strays from purely biological topics into speculations about human behaviour in social life, there is a regrettable tendency for biology to be deployed in the service of upholding the existing order. Thus, for example, the economic and social disadvantages of women compared to men have been ascribed, not to inequities in the economy and society, but to a supposed biologically determined order. Early human hunter-gatherers, according to this idea, were characterized by active, aggressive men who went out hunting, while women stayed at home to look after the children; and what was true of the Stone Age is also true of the twentieth century. In a 1975 article in the *New York Times* magazine, E. O. Wilson, the greatest proponent of sociobiology, remarked: 'This strong bias persists in most agricultural and industrial societies and, on that ground alone, appears to have a genetic origin ... My own guess is that the genetic bias is intense enough to cause a substantial division of labor even in the most free and most egalitarian of future societies.' Wilson's comments are an interpretation and there is no direct evidence to support them. More recent work by anthropologists and palaeontologists has cast serious doubt on the traditional picture of man the hunter, so the supposed archaeological base for the idea is shaky.

A different interpretation might start by citing copious historical evidence on how men have consistently denied women access to basic education. For example, in Britain to this day the quality of school-teaching in mathematics is frequently inferior when the pupils are girls rather than boys. As late as the 1940s, Cambridge University, one of the foremost in Britain, was refusing to confer degrees on women: women might attend classes, but they could not graduate. It is rather hard to see how the degree-awarding policy of Cambridge University could have been in accordance with genetic determinism up to 1950, but have escaped from genetic control since then. Nor, one suspects, would the Cambridge dons be flattered by an explanation of their earlier conduct in terms of Palaeolithic atavism. By the pursuit of such non-genetic analyses, all traces of alleged Stone Age influence are very quickly buried under the palpable weight of past and present day discrimination.

But biological and genetic 'explanations' for social, cultural or economic differences have a perennial fascination. One may expect the arguments put forward by those who seek biological explanations of social behaviour to grow more subtle or at least to sound superficially more convincing as they will be increasingly decked out in a vocabulary of genetic determinism borrowed from and influenced by the successes of the Human Genome Project. One reason for the popularity of such explanations is perhaps their straightforwardness and simplicity: by appealing to a genetic explanation, one can mimic the thought-processes of the atomic physicists and strip away extraneous complications to get down to the fundamentals, the underlying causes. If, by disaggregating the material world down to its elementary particles, the physicists have been able with sensational success to derive general laws and principles governing the operation of the whole inanimate universe, surely such a procedure would yield laws or principles of general applicability to the living world, and to humanity in particular?

At first sight, genes appear attractively simple things with which to deal compared with the messy complexity of human societies with all their arbitrary institutions which have been brought into existence by accidents of history. Bound up in this view is the feeling that genetic explanations appear to shed light on that most elusive of concepts, human nature. As Max Charlesworth pointed out in a paper presented to a symposium on human genetics organized by the Ciba Foundation in 1989, sociobiologists and some geneticists make the claim 'that there is a direct connection between scientific evidence, derived from evolutionary biology and genetics, and the concept of human nature'.

The rise of the environmental movements in many Western societies since the late 1960s also has a bearing on the attractiveness of biological (which almost without exception means genetic) explanations for human conduct. There is a strong feeling that Nature and human nature have grown apart from each other and that they must be brought back together again. H. L. Kaye remarked in his book *The Social Meaning of Modern Biology* that many people feel strongly that biology and society ought to be in step. Theories such as sociobiology, which purport to explain modern social life in terms of underlying biology, appeal to a type of Romanticism, 'a nostalgia for a mythical Ur-harmony between human biological nature and human cultural forms', Kaye believes. Some proponents of the Human Genome Project might protest that such a borrowing of genetic explanations of human traits and behaviour is illegitimate, but on the other hand, a few of the project's foremost proponents have claimed that their science will indeed yield new insight into human nature.

Walter Gilbert, a Nobel Prize-winning geneticist, remarked that when we know the complete human genome we will know what it is to be human. A similar claim was made somewhat more poetically by Robert Sinsheimer, one of the original movers of the Human Genome Project in the United States: 'We seek not in the stars but in our genes for the herald of our fate.'

There is one major difference in the assumptions underlying sociobiology and the Human Genome Project. For sociobiology, the 'iron hand of the genes' is immutable, human nature is fixed. For geneticists, however, knowing the composition of the human genome opens the way to altering it. Sinsheimer, for example, would go further than merely acquiring knowledge about the genetic composition of humanity and would change and 'improve' the human race, and he celebrates the fact that 'A new eugenics has arisen [. . .] one of the most important concepts to arise in the history of mankind. I can think of none with greater long-term implications for the future of our species.' In the previous chapter, several arguments were advanced suggesting that eugenic applications of the new genetics were still some way off in the future. But the risk remains that rather than altering our society, we may employ our new knowledge to alter ourselves.

The Human Genome Project will bring political fall-out too. Some authors, including Stephen Jay Gould in *Ever since Darwin*, have warned that those with vested interests in the social and political *status quo* may try to hijack biology and pretend that the prevailing social order is rooted in biological and genetic nature, and therefore is unalterable. One finds Bertrand Russell thinking about the possibility of illiberal political consequences of modern biology as long ago as 1946: 'The doctrine that all men are born equal and that the differences between adults are due wholly to education was incompatible with [Darwin's] emphasis on congenital differences between members of the same species.' Culturally, throughout its history, humanity has been suspicious of its own diversity and difference has been a pretext for persecution. In our own time, fear of the different has been a central theme of many science fiction novels. John Wyndham's *The Chrysalids* is set among the survivors of a nuclear holocaust where the radiation induces many more mutations but where any manifestation of the increased mutation rate is met with immediate death. But although Wyndham's book was set in a post-nuclear future, its theme reflects the normal behaviour of humanity. Genocide has been regrettably frequent in humanity's history. The motivation has usually been cultural or economic: although the public justification for extirpating the American Indians was their alleged savagery and warlike behaviour, they also happened to occupy prime agricultural land coveted by the white settlers.

Genetics is particularly vulnerable to misuse in providing an allegedly scientific basis for discrimination which is actually founded on cultural or economic conflict. One of the purposes of the study of genetics was to understand the mechanism which provided variation between the individual members of a given population of animals, for it is on this variation that the Darwinian processes of natural selection operate. If regarded too narrowly, genetics as applied to humans appears to stress the differences between members of human society. But in biological terms such differences are advantageous to the overall survival of the species, for natural selection does not act solely to provide for the survival of the fittest individuals in a population. Herbert Spencer went down the wrong track when he coined his phrase about 'survival of the fittest' and the Social Darwinists erred not only in their politics but also in their biology. It is necessary to look at the implications of genetics for the population as a whole and not just the individuals within it. Evolution strikes a balance between immediate fitness and longer-term genetic flexibility. Natural selection in combination with Mendelian patterns of inheritance tends to maintain a degree of genetic flexibility in large populations – there is a clustering round the average, but the extremes of variation are seldom lost. This tendency to maintain diversity provides the raw material upon which natural selection can act, and is a more important source of variation than are the comparatively few beneficial mutations that may occur randomly in the genes. If the environment changes – if a new disease were to break out – then natural selection would nudge the average in a slightly different direction because those who happened to be slightly or wholly resistant to the disease would tend to have a better chance of survival and procreation. For the geneticist, diversity within our species is something to be prized. That of course is a biological judgement and does not necessarily carry any moral connotations, but it does serve to illustrate how some attempts to ground cultural or economic conflicts in the language of biological (genetic) difference are without scientific credibility.

One of the problems in the near future will be to distinguish between some traits which are genetically determined (such as eye colour) and those which are merely genetically influenced. To take a fairly uncontroversial example, even before the turn of the century Francis Galton had established that tall people tend to have tall children. But it is also true that what a child eats will affect its final height. Anyone who visits my native city of Glasgow will be struck by the numbers of short older men. I am considerably taller than my father, and most Glaswegians of my age are taller than the previous generation, partly because we all had better diets in the post-war era than was usual during the years of unemployment and depression experienced by our parents when they were young. Height is thus a mixture of nature (the genetic influence) and nurture. Many other, more controversial, traits are also likely to be the result of an interplay between genes and environment.

Intelligence is clearly so influenced. In 1932, a survey was made of the performance of eleven-year-old Scottish children in IQ tests; the exercise was repeated with 70,000 eleven-year-olds fifteen years later, in 1947. If it can be assumed for the moment that IQ tests have anything to do with intelligence (and this is a controversial assumption), then the intelligence of the boys had increased slightly, while that of the girls had increased significantly over the fifteen-year period. The children in the second cohort were on the average taller than the earlier group, which it seems reasonable to assume was the consequence of better nutrition and of a better home environment generally. So it also seems reasonable to assume that intelligence is affected by improvements in nutrition and home environment.

Equally it seems reasonable to assume that there is some genetic variability in intelligence, and in the late 1980s, the Minnesota twins study – a project which compared the IQ test performance of identical twins who had been reared apart with those who had been reared together – claimed to find just such an effect. There is a risk that over the next decade or so, with the practical applications of the Human Genome Project filtering through into the diagnosis and treatment of human diseases, there may be much less emphasis on the environmental contributions to the make-up of a human being and an almost exclusive concentration on the genetic constitution and on genetically influenced differences.

But even differences that do have a basis in inheritance can be misunderstood. Height and intelligence differ from the simple examples of inheritance. Human beings do not fall into two separate categories of tall and short: there is a smooth continuum of varying height with the majority of people around the average height. The normal variation in human height is clearly not under the control of a single gene, but of a polygenic system. Many genes, of varying dominance or none at all, act together in a complex and as yet unknown fashion to give rise to the genetic component of the height that an adult human attains. Human intelligence is likewise a polygenic trait: there is a continuum of intelligence with the majority of the population around the average. It seems certain that the majority of the most interesting human traits will be polygenic and will not conform to the simplified picture of there being 'a gene for' intelligence or any other trait, which can be inherited in some simple manner. In particular, human behaviour (as opposed to the structural features of our bodies) is so complex, variable and flexible, that although it may be subject to genetic constraints, it is very difficult to make a case for any particular gene or gene system being responsible for a specific pattern of human behaviour.

This brief discussion of polygenic inheritance indicates two things. Appeals to biological (genetic) explanations for human behaviour and human social institutions do not yield simple answers. It is likely to be equally if not even more complicated to unravel the interplay of the different components within a polygenic system and their interaction with the environment as it is to analyse social and political events in social and political terms. The second point is to warn against a possible tendency to seek biological explanations in terms of there being 'a gene' for a particular trait, and not in terms of the complexities of polygenic inheritance. Unfortunately but understandably, the unravelling of the human genetic constitution has proceeded largely along the lines of uncovering aberrant variations in single genes which have a profound effect on the individual carrying the gene. This is unsurprising, for such single-gene traits are the easiest to work on, and science proceeds by first asking the question which is easiest to answer and only then proceeding to the more difficult questions. So, even when all the genes in the human genome have been identified, there will still be many years of work before we understand the complexities of their interaction and the subtle ways in which apparently unrelated genes may actually influence each other. There is a risk that those unfamiliar with the detailed work of piecing together this genetic mosaic may not appreciate that the interactions are ultimately of more significance than the single genes. The lay public may switch off at a halfway point in understanding the genome, and their political leaders (who certainly count as members of the lay public in this context) may formulate policy and legislation on the basis of a half-understanding of the real situation. Unfortunately, some researchers appear to condone or even encourage such superficial approaches to the complexity of human genetics. In 1988, an international team published a study of the patterns of incidence of schizophrenia in Icelandic and British families. The patterns indicated that a gene linked with susceptibility to schizophrenia might lie on chromosome number 5. Schizophrenia is almost certainly a polygenic disease of some sort with several different genes contributing to its occurrence. However, the press release distributed to British newspapers at the time by one of the institutions involved had no room for such subtleties. It stated quite baldly that the gene that causes schizophrenia had been found. Subsequent studies in any case failed to substantiate the link between schizophrenia and chromosome 5 and the researchers have since withdrawn their original findings – quietly and without circulating press releases to the newspapers.

Discussion of how the Human Genome Project will affect our view of ourselves and of our place in the natural world involves psychology rather than technological development. Yet the knowledge which the project will uncover will not be passive; it will change things. Perhaps the most unsettling consequence of the genome project might be the spread of the idea that a human being is no more than the biological expression of the program of instructions encoded in his or her DNA. In most popular accounts of the subject, the computing metaphors are already building up fast (and they are so convenient that the reader will spot many instances of their use throughout this book). But to continue the metaphor, what chance is there for the moral uniqueness of our program as distinct from that of any other creature? Will it be possible to continue to ascribe moral worth to being human when there is a risk that humanity might appear to itself little more than the working-out of a genetic computer program? In the face of the avalanche of genetic information which will be unleashed by the project will we be able to avoid the error, pointed out by the English philosopher Mary Midgley among others, of equating a product with its source? Midgley has dubbed it the 'genetic fallacy' – 'to say that a flower is really only organised dirt' – but it will become increasingly difficult to avoid the temptation of saying that human beings are 'really' only the expression of their genes. The significance of the Human Genome Project is not that it has discovered or invented these ideas. But it will bring these ideas down from the ivory tower, forcing the rest of us to confront them squarely for the first time.

The Human Genome Project is up and running, but only the first few fruits of the research have become available. Even were there no co-ordinated project, the progress of research to mitigate or cure human disease would inevitably encounter moral, social and legal questions. The genome project simply throws them into higher relief and makes them more urgent. Although some scientists may deny it, the advance of biological and genetic research does seem to be raising new moral and ethical questions – or to be recasting the old ones in a different frame, with a different vocabulary, and with a more immediate and more widespread impact. Since no one has been able satisfactorily to answer the 'old' moral questions, it should not be surprising if the new ones too present difficulties.

The existence of the new genetic knowledge will have a profound effect on not only on individuals but on social institutions, which have a history and their own inertia and 'will to live'. Einstein commented, à propos atomic energy, that the nation state and the split atom cannot coexist on the same planet. But when the atom was split, nationalism as a form of social organization did not collapse. Racial, religious, cultural and linguistic divisions continued to exist and to matter more than common humanity. Indeed, these very divisions called the atomic bomb into existence, for there was no scientific reason for the atomic physicists to build a bomb rather than a power-producing nuclear reactor. The potential knowledge of how to build a bomb would always have been there, but the decision to build it and to pursue the necessary research were culturally determined. The episode of the bomb illustrates how society's values can affect, indeed determine, the direction that applied scientific research takes. And there is nothing more applied than medical research.

This raises the fundamental question: 'Who is to decide?' The second half of this book has reviewed many of the decisions taken by societies over the past half century with respect to the application of human genetics. While many people have been helped and much human suffering has been alleviated, it none the less remains true that collectively, society has sometimes dealt very unwisely with this subject. The democracies of the Western industrialized nations have on occasion displayed a surprisingly illiberal streak and have discriminated against those whose genetic constitution (actual or presumed) made them different. How is society to ensure that, with the much more powerful tools becoming available from the Human Genome Project, such actions are not repeated? Ultimately the question is not just one of trying to discern which choices are good and which ones morally unacceptable, but of giving effect to such a decision through legislation, both nationally through parliament, congress or national assembly, and internationally by intergovernmental declarations and conventions through such bodies as the Council of Europe or the World Health Organisation.

Most Western societies are pluralistic, with many different moral and religious traditions within one country. Within the UK, for example, there are sizeable communities embracing religious faiths as different as Islam, Roman Catholicism, Judaism, Anglicanism and Calvinism, while about a third of the population no longer claim any religious faith. In such circumstances, how is consensus to be achieved? It is clear that there are some things which should be prohibited, at least for the time being, until we understand the science and ourselves better. Society needs to demarcate those things which will be permitted and to list those which, for the moment, ought to be forbidden. Ultimately, such decisions have to be taken, in a democracy, by the elected representatives of the people. But politicians are busy individuals, and the debating floor of the national legislature is not a good place in which to rehearse difficult arguments about what is technically possible and morally desirable. There is thus a need for each country to explore these issues in a public but non-political forum where it may be possible to inch towards some sort of consensus, despite the plurality of moral beliefs, or at least to expose the principal differences. Only if such matters are ventilated publicly can politicians hope to avoid being stampeded into hasty legislation by newspaper headlines or the activities of some minority lobby group.

Beyond moral or legal issues, the practical question that the

Human Genome Project sets out to answer is, 'In genetic terms, what is human?' We already know that the answer will come in about 3 billion parts. As the information comes ever faster out of the laboratories, it will be hard to remember that the Human Genome Project set out just to examine human genetics – and that the answers it yields are only partial insights into the human condition. Genes are important, but they are not all-important. The objective of the Human Genome Project is to draw up a new anatomy, a neo-Vesalian basis for the medicine of decades to come. But people are more than their physical anatomy. Our anatomy rules out certain futures, but it does not dictate which of the many possible futures that could be mine will be mine. Our physical and genetical anatomies represent constraints; they do not represent predestination.

Nor is our physical anatomy indispensable to our personal identity. If as a child I had lost both legs in an accident, then it would have severely changed my personal anatomy and would have ruled out many otherwise attainable futures. In some sense I could not have become the person that I am today, but even though my life history would have been different it would still recognizably be the same 'I'. Similarly, I could have a heart and lung transplant, liver and kidneys, bone marrow, and still be me.

So it is with the genetic anatomy. The human genome is just that: human anatomy expressed in genetic terms. There is a fatal temptation to import into biology some of the concepts and habits of thought peculiar to theoretical physics. The physicist searches for an inner simplicity and fundamental structure to the world of forces and matter. This search has led to the concept of the 'fundamental forces' which govern the material world and to the idea that material objects are composed of elementary constituent particles. Once the particles and the forces that govern their interaction have been specified, then the subsequent development of the universe can be predicted, at least in principle – only a shortage of the necessary computational capacity prevents us from doing so in practice. The temptation is to regard genes as fundamental entities in biology in a similar vein to the elementary particles and to think that by specifying an individual's genes, one has somehow set out their life's course. It is a common misperception, of which almost everyone is guilty, to think that the genes somehow encapsulate the inner 'essence' of an individual. Yet this is clearly not true. The DNA of a chimpanzee and that of a human being differ by only 1.6 per cent: it seems incoherent to assert that that 1.6 per cent of DNA is the part which contains the essence of humanity. Assuming it were possible, about 90 per cent of my DNA – the 'junk' DNA which does not code for any genes nor play any part in cell replication – could have been taken out of the fertilized egg from which I developed and, on the basis of present knowledge, it appears that none of us would be any the wiser. Even in those genes which code for proteins it seems difficult to find my inner essence. For example, all that differs between me and someone who suffers from sickle-cell anaemia is one base pair in both copies of the gene which specifies haemoglobin. Had I been born with the sickle-cell mutation then my life would have been very different, yet it is difficult to see how a genetic injury differs in kind from the hypo-thetical case where I sustain an environmental injury by losing both legs – I would not have been changed 'in essence'.

And so at last one comes to a profound irony. By the time it is complete, the Human Genome Project will have cost more than $3 billion and occupied the energies and intellects of thousands of the world's most creative scientists over a period of nearly two decades. In the course of that research, and in the working-out of its implications, much previously intractable human suffering will have been avoided or alleviated. It will have set out a complete genetic blueprint for humanity, uncovering not only the differences between one human and another but also deep underlying similarities between humans and the rest of the living world. Yet at the end, after all that effort, what is perhaps the project's most important function should be to transcend itself and teach us, or to remind those who should not need teaching, that genes and genetics are not the fundamental basis of human life. The project will have been successful if it accomplishes the difficult task of persuading us all to abandon the habits of thinking which have been inherited from the physicists, where the elementary particles are both fundamental essence and ultimate cause. Physicists see the world in terms of entities – things such as material objects or abstract quantum fields – whose ultimate fate is fixed by their initial configuration. The physicist's view of the world is thus timeless: there is no becoming, there is no narrative of events because the future is frozen into the state of the present.

In contrast, life is not a collection of things – whatever I am today is very different from what I was at birth, both in terms of physical composition and, more importantly, in terms of my personal identity. A life has more the character of a story which unfolds and in which all incidents are equally necessary to the story's completeness: one cannot assert with the Emperor Joseph II that *Cosi fan tutte* contains too many notes; neither can one rip a page out of *Hamlet*, claiming that it represents the essence of the human tragedy and that no other pages need to be read. The physicist's universe is in stasis, whereas life is a process – like a flickering candle flame which has a form and consumes energy. But, although life leaves a trace of itself in the form of genetic inheritance when at last the candle has been used up, there is no essence to be found.

This then may be the final challenge posed by the Human Genome Project: to redefine our sense of our own moral worth and to find a way of asserting, in the face of all the technical details of the genetics, that human life is greater than the DNA from which it sprang, that human beings retain a moral value which is irreducible and which transcends the sequence of 3 billion base pairs within the human genome.

INDEX

Abortion, selective, 12, 112–13, 121–2, 123–5, 129–32, 155
Access to information, 10–12, 126, 133
Adenosine deaminase (ADA), 16–23, 69, 151, 152–3
Advisory Council on Science and Technology, 87–8
AIDS, 101, 159
Ajl, Samuel, 86
Alzado, Lyle, 139
Alzheimer's disease, 19, 122
Amino acids, 34–6
Amniocentesis, 47, 111
Animal kingdom, 171, 172–8
Athletes, 138–40
Avery, Oswald, 29

Baltimore, David, 64, 79, 83
Base pairs, 30–32, 66–7, 70
Becker muscular dystrophy, 25
Berg, Paul, 59, 60, 66
Bias, Wilma, 52
Biotechnology, 60–62, 134–51
Blaese, Michael, 16, 20, 22, 23, 151
Blood groups, 47, 52, 174
Bodmer, Sir Walter, 74, 92
Bolivar, Francisco, 61
Bone-marrow transplantation, 19–20, 148–9
Botstein, David, 74, 79–80
Boyer, Herbert, 59–60, 61
Brenner, Sydney, 87, 88, 90
Brewer, George, 149–50
Brock, D. J. H., 115

Caenorhabditis elegans, 90, 94
Callahan, Daniel, 98–9
Cancer: and gene therapy, 22, 38–9; and genetic damage, 37–9; and screening, 122, 124, 125, 128–9
Carter, Jimmy, 132, 156
Caspersson, Torbjörn, 46
Centre d'Etude du Polymorphisme, 90–91
Chargaff, Edwin, 30
Charlesworth, Max, 181

Chase, Martha, 29
Childs, Barton, 49
Chorionic villus sampling, 47, 112
Chromosomes, 41–54, 65–6, 89
Churchill, Winston, 13
Cline, Martin, 23, 153–4
Cloning, gene, 58–62
Codons, 35–6, 70
Cohen, Stanley, 59–60
Cold nuclear fusion, 143–4
Collins, Francis, 26, 95
Colour vision, 44, 50, 128
Comings, David, 98
Commercial issues, 5–6, 8–9, 60–62, 94, 95–6, 100
Complementary DNA, 64–5, 88–9
Contig mapping, 63–4, 86
Creutzfeld-Jacob disease, 135–6
Crick, Francis, 4, 29–30, 34–5, 80
Crossing over, 48–50, 89
Culver, Ken, 16
Cystic fibrosis, 2, 19, 25–8, 47, 154; screening for, 28, 55, 114–20, 123

Darwin, Charles, 168–9, 172–3, 177
Dausset, Jean, 90
Davenport, Charles, 164
Davies, Kay, 24
Davis, Ronald, 74
Delbrück, Max, 56
DeLisi, Charles, 78–9, 82
Denmark, 91
Deparment of Energy (DOE), 77–9, 81, 82, 83, 85
Desferrioxamine, 150
Diabetes, 122, 124
Diagnosis, genetic, see Screening
Discrimination, 171, 178–87; see also Eugenics
Disease(s), 2, 3, 5, 40–41, 45–6; chromosome abnormalities, 42–3, 47–8; and gene transplantation, 2, 16–23; and genetic markers, 23–8; inheritance of, 18–19, 25, 26; and mutation, 35–6; screening for, 28,

38, 55, 97–133; sex-linked, 19, 24, 43–4; *see also individual disorders*

DNA (deoxyribonucleic acid), 20; complementary DNA, 64–5, 88–9; damage to, 37–8; functionless sequences of, 69–71, 88–9; libraries, 63–4; probes, 64–6; and protein synthesis, 33–6; recombinant DNA technology, 54, 59–68; replication of, 31–2, 37; sequencing of, 66–7; structure of, 4, 29–32, 41; studies of, 57–68, 73–5; *see also* Gene(s)

DNA Polymerase I, 67–8

Doll, Sir Richard, 9

Donahue, Roger, 51–2

Down's syndrome, 42, 43, 47–8

Dozy, Andrees, 112

Draaijer, Lisa, 139

Drosophila, 42, 50–51

Duchenne muscular dystophy (DMD), 19, 24–5, 69, 154

Dulbecco, Renato, 77

Dystrophin, 24–5

Education, 129, 130, 163, 180

Edwards, Robert, 8

Einstein, Albert, 176, 187

Erythropoetin, 139–40

Eugenics, 13, 163–5, 179, 182

European Community, 11, 14

Evolution, 89, 98, 99, 168, 172–8, 183

Exons, 69

Eysenck, Hans, 142

Factor VIII, 148

Fleischmann, Martin, 143–4

Ford, Charles, 42

Fost, Norman, 119

France, 90–91, 93

French Anderson, W., 16, 22, 38

Fruit flies, 42, 50–51

Galton, Francis, 163, 184

Gamow, George, 34

Gauthier, Marthe, 42

Gel electrophoresis, 58, 68, 104

Gender determination, 42–3, 173–4

Gene(s): cloning, 58–62; mapping, 41–54, 57; markers, 23–8, 75, 85–6; number of, 40, 70; probes, 64–6; and proteins, 17–18, 20, 33–6, 69–70; regulation, 36–7; splicing, 20; transplantation, 2, 16, 20–23, 38–9, 151–65

Genentech, 61–2, 137

Genethon, 90–91

Genetic code, 35–6

Genome, human, 1–2; studies of, 40–54,

55–68; *see also* DNA; Gene(s)

Gilbert, Walter, 66, 81, 182

Glucose-6-phosphate dehydrogenase (G6PD), 49–50, 99, 102, 127

Goodfellow, Peter, 43

Goodman, Robert, 129–30

Gould, Stephen Jay, 182

Green, Howard, 52–3

Growth hormone, 61–2, 135–9

Gusella, James, 75

Guthrie, Woody, 72–3

Haemoglobin, 17–18, 36, 89, 102–5, 153

Haemophilia, 44, 148

Harris, John, 161–2

Heart disease, 19, 122, 124–5

Herrick, James, 104

Hershey, Alfred, 29

Hood, Leroy, 66

Horner, Friedrich, 44

Human Genome Organisation, 6, 92

Human Genome Project, 1–2, 28, 41, 83–96; commercial aspects of, 5–6, 8–9, 94, 95–6, 143, 154; genesis of, 76–83; health care systems and, 121, moral aspects of, 3–4, 6–15, 98–9, 135, 136, 169–91

Human growth hormone, 61–2, 135–9

Hume, David, 176

Huntingdon's chorea, 19, 47, 72–3, 75, 122–4

Huxley, T. H., 178–9

Hybridization, 64–5

in situ hybridization, 65

Ingram, Vernon, 104

Inheritance, 18; of disease, 18–19, 25, 26; of genetic damage, 37–8; and germ-line therapy, 151–2, 158–65; and recombination, 48–50; and screening, 28, 38, 55, 97–133; studies of, 40–54

Insulin, 62

Insurance, 10–11, 107, 115, 120–121, 125–7, 147, 155

Intelligence, 3–4, 12, 13, 129–33, 140–143, 144–6, 163–4, 184–5

Introns, 69–70, 88–9

Itakura, Keiichi, 61

Itano, Harvey, 104

Jackson, David, 59

Jackson, Michael, 143

Jacobs, Pat, 42

Japan, 91–2

Jensen, Arthur, 141–2

Jo Hin Tjio, 42

Judson, Horace Freeland, 86

Kan, Y. W., 112
Karyotyping, 46–8
Kaye, H. L., 181
Kirschstein, Ruth, 83
Klinefelter syndrome, 42–3
Kopelman, Loretta, 107, 108
Kornberg, Arthur, 67
Koshland, Daniel, 80
Kuhn, Thomas, 167
Kunkel, Louis, 24

Lejeune, Jerome, 42
Levan, Albert, 42
Linkage maps, 53
Lovell-Badge, Robin, 43
Luria, Salvatore, 56

McCarty, Maclyn, 29
McKusick, Victor, 40, 50, 92
MacLeod, Colin, 29
Malaria, 45, 97–9, 103–5, 161
Maniatis, Tom, 62–3
Marks, Joe, 80
Maxam, Allan, 66
Mayall, Ed, 55, 67–8
Medawar, Sir Peter, 98, 163
Medical Research Council (MRC), 87, 88,
   115
Mendel, Gregor, 168–9
Midgley, Mary, 187
Migeon, Barbara, 53
Miller, C. S., 53
Modell, Bernadette, 113
Monaco, Tony, 24
Moore, G. E., 176
Moore, John, 12
Moral issues, 3–4, 6–15, 166–91; in
   biotechnology, 134–51; in gene
   transplantation, 22–3, 152, 156, 159–65; in
   screening, 98–133
Morgan, Thomas Hunt, 50–51
Murray, J. M., 24
Murray, Robert, 107–8
Muscular dystrophy, 2, 19, 24–5, 69, 154
Mutation(s), 35–6

National Center for Human Genome
   Research, 5–6, 83, 94–5
National Institutes of Health (NIH), 16,
   20–21, 110, 115, 156, 157–8; and Human
   Genome Project, 79, 81, 82–3, 85, 92–3,
   94–5
National Research Council, 81–2
Nazi Germany, 13, 164, 179
Nematode worm, 90, 94
Newton, Sir Isaac, 168, 170

Nuffield Foundation Council on Bioethics,
   14

Olson, Maynard, 64
Oppenheimer, J. Robert, 4
Orgel, Leslie, 32

Patents, 6, 10, 12, 93–4, 95–6
Pauling, Linus, 104, 106
Perutz, Max, 148
Peters, Keith, 88
Phenylketonuria, 105
Pines, Maya, 53–4
Polygenic disorders, 19, 122, 124–5
Polymerase chain reaction, 67–8
Pons, Stanley, 143–4
Porter, Ian, 50
President's Commission for the Study of
   Ethical Problems in Medicine and
   Biomedical and Behavioral Research,
   14–15, 109–10, 156
Privacy, 10–12, 121, 126, 133
Protein(s), 17–18, 20, 29; analysis, 23;
   synthesis, 33–6, 69–70; see also
   Biotechnology
Pseudogenes, 89

Rachels, James, 173, 176–8
Recombinant DNA technology, 54, 59–68
Recombination, 48–50, 89
Renwick, James, 52
Restriction enzymes, 58–9, 62–3, 73–4
Restriction fragment length polymorphism,
   73–5
Reverse genetics, 23–8, 75
Reverse transcriptase, 64–5
Rights, individual, 10–12, 109, 126, 133
RIKEN project, 91–2
RNA (ribonucleic acid), 32–3; and cDNA,
   64–5, 88–9; and protein synthesis, 35–6
Roberts, Richard, 69
Rosenberg, Steven, 16, 22, 38–9
'Rothley report', 14
Ruddle, Frank, 53
Russell, Bertrand, 173, 182

Sanger, Fred, 66
Schizophrenia, 186
Schulze, J., 50
Screening, 28, 38, 55, 97–133
Sequencing, DNA, 66–7
Severe Combined Immune Deficiency
   (SCID), 16–17, 23, 148–9
Sex chromosome, 42–4
Sex-linked disorders, 19, 24, 43–4
Sharp, Philip, 69

Shows, Thomas, 54
Sickle-cell anaemia, 2, 19, 23, 35–6, 45, 98, 99–100, 102–13, 148, 149, 150, 159–61
Single-gene disorders, 18–19, 25
Sinsheimer, Robert, 76, 182
Skolnick, Mark, 74
Smith, Hamilton, 58
Social Darwinism, 179, 183
Sociobiology, 179–82
Solomon, Ellen, 74
Somatic cell hybrids, 52–3, 65
Somatostatin, 61, 62
Somatotropin, 61–2, 135–9
Spencer, Herbert, 178, 183
Sports, 138–40
State/society, 9, 11, 131–3, 146, 155–6, 162–5, 171, 178–88
Sterilization, enforced, 13, 155–6, 164
Sulston, John, 90, 94
Symons, Robert, 59

Tay-Sachs disease, 112, 113, 114, 121–2, 177–8
Temin, Howard, 64
Thalassaemia(s), 2, 19, 45–6, 69, 100, 102, 103, 104–5, 113–14, 150, 153–4, 159–61
Thatcher, Margaret, 87, 88
Thymidine kinase, 53
Transplantation: bone-marrow, 19–20, 148–9; gene, 2, 16, 20–23, 38–9, 151–65.
Tsui, Lap-Chee, 26
Turner's syndrome, 42

United Kingdom genome project, 87–90, 93–4

Walsh, James, 80
Warnock, Baroness Mary, 159, 162–3
Waterston, Bob, 90, 94
Watson, James, 4, 29–30, 56, 80; and Human Genome Project, 4–5, 6, 7, 28, 81, 82, 83–4, 93, 94–6
Weatherall, Sir David, 8, 102, 124
Weinberg, Robert, 80
Weiss, Mary, 52–3
Wexler, Nancy, 72, 75, 122–3, 126–7
White, Ray, 26, 74–5
Wilfond, Benjamin, 119
Wilkins, Maurice, 30
Williams, Bernard, 13
Williamson, Bob, 26, 55, 116, 117–18
Wilson, E. B., 44
Wilson, E. O., 180
Wilson, Peter, 11
Woese, Carl, 175
Workplace, 127–9
'Wrongful life' suits, 121–2
Wyndham, John (The Chrysalids), 182–3
Wyngaarden, James, 83

Yeast artificial chromosomes, 64

Zech, Lore, 46
Zinder, Norton, 76–7, 83